RIVER PUBLISHERS SERIES IN INFORMATION SCIENCE AND TECHNOLOGY

Series Editors

K. C. CHEN
National Taiwan University
Taipei, Taiwan
and
University of South Florida, USA

SANDEEP SHUKLA
Virginia Tech
USA
and
Indian Institute of Technology Kanpur, India

Indexing: All books published in this series are submitted to the Web of Science Book Citation Index (BkCI), to CrossRef and to Google Scholar.

The "River Publishers Series in Information Science and Technology" covers research which ushers the 21st Century into an Internet and multimedia era. Multimedia means the theory and application of filtering, coding, estimating, analyzing, detecting and recognizing, synthesizing, classifying, recording, and reproducing signals by digital and/or analog devices or techniques, while the scope of "signal" includes audio, video, speech, image, musical, multimedia, data/content, geophysical, sonar/radar, bio/medical, sensation, etc. Networking suggests transportation of such multimedia contents among nodes in communication and/or computer networks, to facilitate the ultimate Internet.

Theory, technologies, protocols and standards, applications/services, practice and implementation of wired/wireless networking are all within the scope of this series. Based on network and communication science, we further extend the scope for 21st Century life through the knowledge in robotics, machine learning, embedded systems, cognitive science, pattern recognition, quantum/biological/molecular computation and information processing, biology, ecology, social science and economics, user behaviors and interface, and applications to health and society advance.

Books published in the series include research monographs, edited volumes, handbooks and textbooks. The books provide professionals, researchers, educators, and advanced students in the field with an invaluable insight into the latest research and developments.

Topics covered in the series include, but are by no means restricted to the following:

- Communication/Computer Networking Technologies and Applications
- Queuing Theory
- Optimization
- Operation Research
- Stochastic Processes
- Information Theory
- Multimedia/Speech/Video Processing
- Computation and Information Processing
- Machine Intelligence
- Cognitive Science and Brian Science
- Embedded Systems
- Computer Architectures
- Reconfigurable Computing
- Cyber Security

For a list of other books in this series, visit www.riverpublishers.com

High-Performance and
Time-Predictable
Embedded Computing

High-Performance and Time-Predictable Embedded Computing

Editors

Luís Miguel Pinho

CISTER Research Centre, Polytechnic Institute of Porto, Portugal

Eduardo Quiñones

Barcelona Supercomputing Center, Spain

Marko Bertogna

University of Modena and Reggio Emilia, Italy

Andrea Marongiu

Swiss Federal Institute of Technology Zurich, Switzerland

Vincent Nélis

CISTER Research Centre, Polytechnic Institute of Porto, Portugal

Paolo Gai

Evidence Srl, Italy

Juan Sancho

ATOS, Spain

LONDON AND NEW YORK

Published 2018 by River Publishers
River Publishers
Alsbjergvej 10, 9260 Gistrup, Denmark
www.riverpublishers.com

Distributed exclusively by Routledge
4 Park Square, Milton Park, Abingdon, Oxon OX14 4RN
605 Third Avenue, New York, NY 10017, USA

High-Performance and Time-Predictable Embedded Computing / by Luis Miguel Pinho, Eduardo Quinones, Marko Bertogna, Andrea Marongiu, Vincent Nelis, Paolo Gai, Juan Sancho.

Routledge is an imprint of the Taylor & Francis Group, an informa business

ISBN 978-87-93609-69-3 (print)

While every effort is made to provide dependable information, the publisher, authors, and editors cannot be held responsible for any errors or omissions.

Contents

4 Mapping, Scheduling, and Schedulability Analysis   **63**

Paolo Burgio, Marko Bertogna, Alessandra Melani,
Eduardo Quiñones and Maria A. Serrano

Preface

Nowadays, the prevalence of electronic and computing systems in our lives is so ubiquitous that it would not be far-fetched to state that we live in a cyber-physical world dominated by computer systems. Examples include pacemakers implanted within the human body to regulate and monitor heartbeats, cars and airplanes transporting us, smart grids, and traffic management.

All these systems demand more and more computational performance to process large amounts of data from multiple data sources, and some of them with guaranteed processing response times; in other words, systems required to deliver their results within pre-defined (and sometimes extremely short) time bounds. This timing aspect is vital for systems like planes, cars, business monitoring, e-trading, etc. Examples can be found in intelligent transportation systems for fuel consumption reduction in cities or railways, or autonomous driving of vehicles. All these systems require processing and actuation based on large amounts of data coming from real-time sensor information.

As a result, the computer electronic devices which these systems depend on are constantly required to become more and more powerful and reliable, while remaining affordable. In order to cope with such performance requirements, chip designers have recently started producing chips containing multiple processing units, the so-called multi-core processors, effectively integrating multiple computers within a single chip, and more recently the many-core processors, with dozens or hundreds of cores, interconnected with complex networks on chip. This radical shift in the chip design paved the way for parallel computing: rather than processing the data sequentially, the cooperation of multiple processing elements within the same chip allows systems to be executed concurrently, in parallel.

Unfortunately, the parallelization of the computing activities brought up many challenges, because it affects the timing behavior of the systems as well as the entire way people think and design computers: from the design of the hardware architecture, through the operating system up to the conceptualization of the end-user application. Therefore, although many-core processors are promising candidates to improve the responsiveness of these systems, the interactions that the different computing elements may have within the chip can seriously affect the performance opportunities brought by parallel execution. Moreover, providing timing guarantees becomes harder, because the timing behavior of the system running within a many-core processor depends on interactions that are most of the time not known by the system designer. This makes system analysts struggle in trying to provide timing guarantees for such platforms. Finally, most of the optimization mechanisms buried deep inside the chip are geared only to increase performance and execution speed rather than providing predictable time behavior.

These challenges need to be addressed by introducing predictability in the vertical stack from high-level programming models to operating systems, together with new offline analysis techniques. This book covers the main techniques to enable performance and predictability of embedded applications. The book starts with an overview of some of the current many-core embedded platforms, and then addresses how to support predictability and performance in different aspects of computation: a predictable parallel programming model, the mapping and scheduling of real-time parallel computation, the timing analysis of parallel code, as well as the techniques to support predictability in parallel runtimes and operating systems.

The work reflected in this book was done in the scope of the European project P-SOCRATES, funded under the FP7 framework program of the European Commission. The project website (www.p-socrates.eu), provides further detailed information on the techniques presented here. Moreover, a reference implementation of the methodologies and tools was released as the UpScale Software Development Kit (http://www.upscale-sdk.com).

Luís Miguel Pinho
Eduardo Quiñones
Marko Bertogna
Andrea Marongiu
Vincent Nélis
Paolo Gai
Juan Sancho

February 2018

List of Contributors

Alessandra Melani, *University of Modena and Reggio Emilia, Italy*

Andrea Marongiu, *Swiss Federal Institute of Technology in Zürich (ETHZ), Switzerland; and University of Bologna, Italy*

Bruno Morelli, *Evidence SRL, Italy*

Claudio Scordino, *Evidence SRL, Italy*

Eduardo Quiñones, *Barcelona Supercomputing Center (BSC), Spain*

Errico Guidieri, *Evidence SRL, Italy*

Giuseppe Tagliavini, *University of Bologna, Italy*

Juan Sancho, *ATOS, Spain*

Luís Miguel Pinho, *CISTER Research Centre, Polytechnic Institute of Porto, Portugal*

Maria A. Serrano, *Barcelona Supercomputing Center (BSC), Spain*

Marko Bertogna, *University of Modena and Reggio Emilia, Italy*

Paolo Burgio, *University of Modena and Reggio Emilia, Italy*

Paolo Gai, *Evidence SRL, Italy*

Patrick Meumeu Yomsi, *CISTER Research Centre, Polytechnic Institute of Porto, Portugal*

Sara Royuela, *Barcelona Supercomputing Center (BSC), Spain*

Vincent Nélis, *CISTER Research Centre, Polytechnic Institute of Porto, Portugal*

List of Figures

List of Tables

List of Abbreviations

ADEOS	Adaptive Domain Environment for Operating Systems, a patch used by RTAI and Xenomai
ALU	Arithmetic logic unit
API	Application Programming Interface
APIC	Programmable timer on x86 machines
AUTOSAR	International consortium that defines automotive API
BCET	Best-Case Execution Time
BFS	Breadth First Scheduling
BOTS	Barcelona OpenMP task suite
BSD	Berkeley software license
CBS	Constant Bandwidth Server, a scheduling algorithm
CFG	Control Flow Graph
CMP	chip multi-processor
COTS	Commercial Off-The-Shelf
CPU	central processing unit
CUDA	Compute Unified Device Architecture
DAG	Direct Acyclic Graph
DDR	double data rate
DMA	direct memory access (engine)
DRAM	dynamic random-access memory
DSP	digital signal processor
DSU	debug system unit
EDF	Earliest Deadline First Scheduler, a scheduling algorithm
ELF	binary format that contain executables
eTDG	Extended Task Dependency Graph
EVT	Extreme Value Theory
FIFO	First-In First-Out
FLOPS	floating-point operations per second
FPGA	Field-Programmable Gate Array
GPGPU	General Purpose Graphics Processing Unit
GPL	General Public License

GPOS	General purpose Operating System
GPU	Graphics Processing Unit
GRUB	Greedy Reclamation of Unused Bandwidth, a scheduling algorithm
HAL	Hardware Abstraction Layer
HMI	Human Machine Interface
HPC	High Performance Computing
HRT	High Resolution Timers
ID	Identifier
IEC	International Electrotechnical Commission
IG	Interference Generator
IID	Independent and Identically Distributed
ILP	instruction-level parallelism
IRQ	Hardware Interrupt
ISA	Instruction Set Architecture
LHC	Large Hadron Collider
LIFO	Last-In First-Out
LL-RTE	Low-level runtime environment
LRU	Least recently used
MBPTA	Measurement-Based Probabilistic Timing Analysis
MCU	Microcontroller Unit
MEET	Maximum Extrinsic Execution Time
MIET	Maximum Intrinsic Execution Time
MPPA	Multi Purpose Processor Array
NOC	network-on-chip
NOP	No Operation
NUCA	non-uniform cache architecture
OpenCL	Open Computing Language
OpenMP	Open multi processing (programming model)
OpenMP DAG	OpenMP Direct Acyclic Graph
OS	Operating system
PCFG	Parallel Control Flow Graph
PCIe	peripheral component interconnect express
PE	processing element
PLRU	Pseudo-LRU
POSIX	Portable Operating System Interface for UNIX
PPC	PowerPC
P-SOCRATES	Parallel Software Framework for Time-Critical Many-core Systems

pWCET	Probabilistic Worst-Case Execution Time
RCU	Linux Read-Copy-Update technique
RM	Resource manager
RT	Real time
RT task	Real-Time task
RTE	Runtime environment
RTOS	Real time operative system
SDK	Software Development Kit
SMP	Symmetric Multi Processor
SMT	simultaneous multi-threading
SoC	System-on-Chip
SOM	system-on-module
SP	Stack pointer
TBB	Thread Building Blocks
TC	Task context
TDG	Task Dependency Graph
TDMA	Time Division Multiple Access
TLB	translation lookaside buffer
TLP	Thread-Level Parallelism
TSC	Task Scheduling Constraint
TSP	Task Scheduling Point
VLIW	Very Large Instruction Word
VPU	vector processing unit
WCET	Worst Case Execution Time
WFS	Work First Scheduling

1

Introduction

Luís Miguel Pinho[1], Eduardo Quiñones[2], Marko Bertogna[3], Andrea Marongiu[4], Vincent Nélis[1], Paolo Gai[5] and Juan Sancho[6]

[1]CISTER Research Centre, Polytechnic Institute of Porto, Portugal
[2]Barcelona Supercomputing Center (BSC), Spain
[3]University of Modena and Reggio Emilia, Italy
[4]Swiss Federal Institute of Technology in Zurich (ETHZ), Switzerland; and University of Bologna, Italy
[5]Evidence SRL, Italy
[6]ATOS, Spain

This chapter provides an overview of the book theme, motivating the need for high-performance and time-predictable embedded computing. It describes the challenges introduced by the need for time-predictability on the one hand, and high-performance on the other, discussing on a high level how these contradictory requirements can be simultaneously supported.

1.1 Introduction

High-performance computing has been for a long time the realm of a specific community within academia and specialized industries; in particular those targeting demanding analytics and simulations applications that require processing massive amounts of data. In a similar way, embedded computing has also focused mainly on specific systems with specialized and fixed functionalities and for which timing requirements were considered as much more important than performance requirements. However, with the ever-increasing availability of more powerful processing platforms, alongside affordable and scalable software solutions, both high-performance and embedded computing are extending to other sectors and application domains.

The demand for increased computational performance is currently widespread and is even more challenging when large amounts of data need to be processed, from multiple data sources, with guaranteed processing response times. Although many systems focus on performance and handling large volumes of streaming data (with throughput and latency requirements), many application domains require real-time behavior [1–6] and challenge the computing capability of current technologies. Some examples are:

- In cyber-physical systems, ranging from automotive and aircrafts, to smart grids and traffic management, computing systems are embedded in a physical environment and their behavior obeys the technical rules dictated by this environment. Typically, they have to cope with the timing requirements imposed by the embedding domain. In the Large Hadron Collider (LHC) in CERN, beam collisions occur every 25 ns, which produce up to 40 million events per second. All these events are pipelined with the objective of distinguishing between interesting and non-interesting events to reduce the number of events to be processed to a few hundreds [7]. Similarly, bridges are monitored in real-time [8] with information collected from more than 10,000 sensors processed every 8 ms, managing responses to natural disasters, maintaining bridge structure, and estimating the extent of structural fatigue. Another interesting application is in intelligent transportation systems, where systems are developed to allow for fuel consumption reduction of railway systems, managing throttle positions, elaborating big amounts of data and sensor information, such as train horsepower, weight, prevailing wind, weather, traffic, etc. [9].
- In the banking/financial markets, computing systems process large amounts of real-time stock information in order to detect time-dependent patterns, automatically triggering operations in a very specific and tight timeframe when some pre-defined patterns occur. Automated algorithmic trading programs now buy and sell millions of dollars of shares time-sliced into orders separated by 1 ms. Reducing the latency by 1 ms can be worth up to $100 million a year to a leading trading house. The aim is to cut microseconds off the latency in which these systems can reach to momentary variations in share prices [10].
- In industry, computing systems monitor business processes based on the capability to understand and process real-time sensor data from the factory-floor and throughout the whole value chain, with Radio Frequency Identification (RFID) components in order to optimize both the production and logistics processes [11].

The underlying commonality of the systems described above is that they are time-critical (whether business-critical or mission-critical, it is necessary to fulfill specific timing requirements) and with high-performance requirements. In other words, for such systems, the correctness of the result is dependent on both performance and timing requirements, and meeting those is critical to the functioning of the system. In this context, it is essential to guarantee the timing predictability of the performed computations, meaning that arguments and analyses are needed to be able to make arguments of correctness, e.g., performing the required computations within well-specified bounds.

1.1.1 The Convergence of High-performance and Embedded Computing Domains

Until now, trends in high-performance and embedded computing domains have been running in opposite directions. On one side, high-performance computing (HPC) systems are traditionally designed to make the common case as fast as possible, without concerning themselves with the timing behavior (in terms of execution time) of the *not-so-often cases*. As a result, the techniques developed for HPC are based on complex hardware and software structures that make any reliable timing bound almost impossible to derive. On the other side, real-time embedded systems are typically designed to provide energy-efficient and predictable solutions, without heavy performance requirements. Instead of *fast* response times, they aim at having *deterministically bounded response times*, in order to guarantee that deadlines are met. For this reason, these systems are typically based on simple hardware architectures, using fixed-function hardware accelerators that are strongly coupled with the application domain.

In the last years, the above design choices are being questioned by the irruption of multi-core processors in both computing markets. The huge computational necessities to satisfy the performance requirements of HPC systems and the related exponential increments of power requirements (typically referred to as the power wall) exceeded the technological limits of classic single-core architectures. For these reasons, the main hardware manufacturers are offering an increasing number of computing platforms integrating multiple cores within a chip, contributing to an unprecedented phenomenon sometimes referred to as "the multi-core revolution." Multi-core processors provide better energy efficiency and performance-per-cost ratio, while improving application performance by exploiting thread-level parallelism (TLP). Applications are split into multiple tasks that run in parallel on different cores, extending to the *multi-core* system level an important challenge already faced by HPC designers at *multi-processor* system level: parallelization.

In the embedded systems domain, the necessity to develop more flexible and powerful systems (e.g., from fixed-function phones to smart phones and tablets) have pushed the embedded market in the same direction. That is, multi-cores are increasingly considered as the solution to cope with performance and cost requirements [12], as they allow scheduling multiple application services on the same processor, hence maximizing the hardware utilization while reducing cost, size, weight, and power requirements. However, real-time embedded applications with time-criticality requirements are still executed on simple architectures that are able to guarantee a predictable execution pattern while avoiding the appearance of timing anomalies [13]. This makes real-time embedded platforms still relying on either single-core or simple multi-core CPUs, integrated with fix-function hardware accelerators into the same chip: the so-called System-on-Chip (SoC).

The needs for energy-efficiency (in the HPC domain) and for flexibility (in the embedded computing domain), coming along with Moore's law, greedy demand for performance, and the advancements in the semiconductor technology, have progressively paved the way for the introduction of "*many-core*" systems, i.e., multi-core chips containing a high number of cores (tens to hundreds) in both domains. Examples of many-core architectures are described in the next chapter.

The introduction of many-core systems has set up an interesting trend wherein both the HPC and the real-time embedded domains converge towards similar objectives and requirements. Many-core computing fabrics are being integrated with general-purpose multi-core processors to provide a heterogeneous architectural harness that eases the integration of previously hardwired accelerators into more flexible software solutions. In recent years, the HPC computing domain has seen the emergence of accelerated heterogeneous architectures, most notably multi-core processors integrated with General Purpose Graphic Processing Units (GPGPU), because GPGPUs are a flexible and programmable accelerator for data parallel computations. Similarly, in the real-time embedded domain, the Kalray Multi-Purpose Processor Array (MPPA), which includes clusters of quad-core CPUs coupled with many-core computing clusters. In both cases, the many-core fabric acts as a programmable accelerator. More recently, the Field-Programmable Gate Array (FPGA) has been used as a flexible accelerator fabric, complementing the above.

In this current trend, challenges that were previously specific to each computing domain, start to be common to both domains (including energy-efficiency, parallelization, compilation, and software programming) and are magnified by the ubiquity of many-cores and heterogeneity across the whole computing spectrum. In that context, cross-fertilization of expertise from both computing domains is *mandatory*.

1.1.2 Parallelization Challenge

Needless to say that many industries with both high-performance and real-time requirements are eager to benefit from the immense computing capabilities offered by these new many-core embedded designs. However, these industries are also highly unprepared for shifting their earlier system designs to cope with this new technology, mainly because such a shift requires adapting the applications, operating systems, and programming models in order to exploit the capabilities of many-core embedded computing systems. On one hand, neither have many-core embedded processors, such as the MPPA, been designed to be used in the HPC domain, nor have HPC techniques been designed to apply embedded technology. On the other hand, real-time methods to determine the timing behavior of an embedded system are not prepared to be directly applied to the HPC domain and these platforms, leading to a number of significant challenges.

On one side, different parallel programming models and multiprocessor operating systems have been proposed and are increasingly being adopted in today's HPC computing systems. In recent years, the emergence of accelerated heterogeneous architectures such as GPGPUs have introduced parallel programming models such as OpenCL [14], the currently dominant open standard for parallel programming of heterogeneous systems, or CUDA [15], the dominant proprietary framework of NVIDIA. Unfortunately, they are not easily applicable to systems with real-time requirements, since, by nature, many-core architectures are designed to integrate as much functionality as possible into a single chip. Hence, they inherently share out as many resources as possible amongst the cores, which heavily impacts the ability to providing timing guarantees.

On the other side, the embedded computing domain world has always seen plenty of application-specific accelerators with custom architectures, manually tuning applications to achieve predictable performance. Such types of solutions have limited flexibility, complicating the development of embedded systems. Commercial off-the-shelf (COTS) components based on many-core architectures are likely to dominate the embedded computing market in the near future, even if complemented with custom function-specific accelerators. As a result, migrating real-time applications to many-core execution models with predictable performance requires a complete redesign of current software architectures. Real-time embedded application developers will therefore either need to adapt their programming practices and operating systems to future many-core components, or they will need to content themselves with stagnating execution speeds and reduced functionalities, relegated to niche markets using obsolete hardware components.

This new trend in the manufacturing technology and the industrial need for enhanced computing capabilities and flexible heterogeneous programming solutions of accelerators for predictable parallel computations bring to the forefront important challenges for which solutions are urgently needed. This book outlines how to bring together next-generation many-core accelerators from the embedded computing domain with the programmability of many-core accelerators from the HPC computing domain, supporting this with real-time methodologies to provide time predictability and high-performance.

1.2 The P-SOCRATES Project

The work described in this book was performed in the scope of the European project P-SOCRATES (Parallel Software Framework for Time-Critical Many-core Systems)[1], funded under the FP7 framework program of the European Commission. The project, finished in December 2016, aimed to allow applications with high-performance and real-time requirements to fully exploit the huge performance opportunities brought by the most advanced COTS many-core embedded processors, whilst ensuring predictable performance of applications (Figure 1.1). The project consortium included Instituto Superior de Engenharia do Porto (coordinator), Portugal, the Barcelona Supercomputing Centre, Spain, the University of Modena and Reggio Emilia, Italy, the Swiss Federal Institute of Technology Zurich, Switzerland, Evidence SRL, Italy, Active Technologies SRL, Italy and ATOS, Spain.

P-SOCRATES focused on combining techniques from different domains: the newest high-performance software techniques for exploiting task parallelism, the most advanced mapping and scheduling methodologies and timing

---

[1]htttp://www.p-socrates.eu

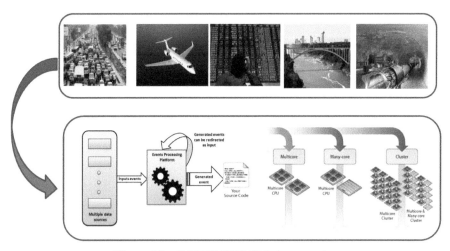

Figure 1.1   P-SOCRATES Global perspective.

and schedulability analysis techniques used in real-time embedded systems, and the low-energy many-core platforms of the embedded domain. This allowed taking important steps towards the convergence of HPC and real-time and embedded domains (Figure 1.2), providing predictable performance to HPC systems and increasing performance of real-time embedded systems.

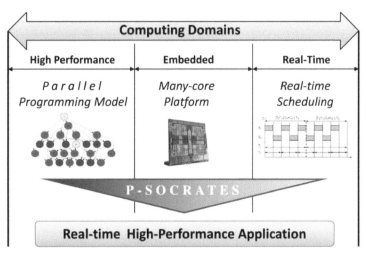

Figure 1.2   P-SOCRATES combines high-performance parallel programming models, high-end embedded many-core platforms and real-time systems technology.

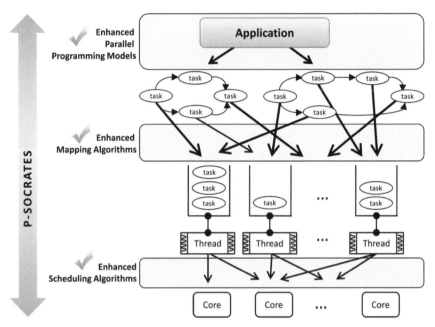

Figure 1.3   Vertical stack of application decomposition.

P-SOCRATES developed a complete and coherent software system stack, able to bridge the gap between the application design with both high-performance and real-time requirements, and the hardware platform, a many-core embedded processor. The project provided a *new framework* to combine real-time embedded mapping and scheduling techniques with high-performance parallel programming models and associated tools, able to express parallelization of applications. The programming model used was based on the state-of-the-art OpenMP specification.

The software stack (shown in Figure 1.3) is able to extract a task-dependency graph from the application, statically or dynamically mapping these tasks to the threads of the operating system, which then dynamically schedules them on the many-core platform.

1.3 Challenges Addressed in This Book

1.3.1 Compiler Analysis of Parallel Programs

In order to enable predictable parallel performance to be analyzed, it is required that the application parallel graph is known, with control- and data-flow information needed for the analysis of the timing behavior of the parallel program. The extraction of this information should be as automatic as possible, to release the programmer from the burden of needing to understand the exact hardware details.

Chapter 3 addresses this challenge by presenting how this information can be obtained from the OpenMP tasking model, and how this information can be used to derive the timing properties of an application parallelized using this model.

1.3.2 Predictable Scheduling of Parallel Tasks on Many-core Systems

To be able to derive guarantees on the correct timing execution of parallel programs, it is required to provide appropriate mapping and scheduling algorithms of parallel computation in many-core platforms, together with deriving the associated offline analysis that enable determining if applications will meet their deadlines.

The challenge of real-time scheduling and schedulability analysis of parallel code is discussed in Chapter 4, which provides the substantial advances that the project has performed in the real-time scheduling and schedulability analysis of parallel graphs, using different scheduling models.

1.3.3 Methodology for Measurement-based Timing Analysis

The use of multi- and many-core platforms considerably challenges approaches for real-time timing analysis, required to determine worst-case execution time of the application code. In fact, the analysis of code execution time is considerably complex due to the interaction and conflicts between the multiple cores utilizing the same hardware resources (e.g., bus, memory, network).

Chapter 5 investigates the different available methods to perform this timing analysis in a many-core setting. After weighing the advantages and disadvantages of each technique, a new methodology is presented based on runtime measurements to derive worst-case estimates.

1.3.4 Optimized OpenMP Tasking Runtime System

The methodology presented in Chapters 3 to 5 of this book relies on the parallel computing abstraction provided by the OpenMP tasking model, and its conceptual similarities to the Direct Acyclic Graph (DAG) model, to achieve predictable task scheduling, requiring an efficient runtime support. However, a space- and performance-efficient design of a tasking run-time environment targeting a many-core system-on-chip is a challenging task, as embedded parallel applications typically exhibit very fine-grained parallelisms.

For that purpose, Chapter 6 presents the design and implementation of an OpenMP tasking run-time environment with very low time and space overheads, which is able to support the approach of the book.

1.3.5 Real-time Operating Systems

The run-time environment of Chapter 6 requires the underlying support of a Real-Time Operating System (RTOS) for many-core architectures. This operating system needs to both be able to execute multi-threaded applications in multiple cores, and also efficiently support a limited pre-emptive model, where threads are only pre-empted at the boundaries of OpenMP tasks.

Chapter 7 presents the re-design and re-implementation of the ERIKA Enterprise RTOS, aiming at an efficient execution on this kind of platforms. The new version of the RTOS allows us to share a single binary kernel image across several cores of the platform, reducing the overall memory consumption, and includes the new limited pre-emptive model.

1.4 The UpScale SDK

An outcome of the P-SOCRATES project was a complete and coherent software framework for applications with high-performance and real-time requirements in COTS many-core embedded processors. This software framework was publicly released under the brand of the UpScale SDK (Software Development Kit)[2]. The UpScale SDK includes the tools to manage the application compilation process, its timing analysis and its execution (Figure 1.4):

- *Compiler flow*. This flow has a twofold objective: (i) to guide the process to generate the binary that will execute on the many-core architecture and (ii) to generate the application DAG used for the timing analysis and run-time components.
- *Analysis flow*. This flow is in charge of deriving timing guarantees of the parallel execution considering execution time traces of the application running on the many-core platform and incorporated in the DAG. Timing

---

[2]http://www.upscale-sdk.com

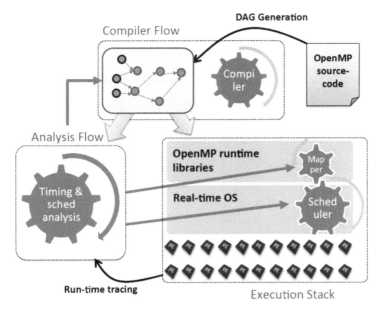

Figure 1.4 The UpScale SDK.

guarantees are derived by means of execution time bounds and a static scheduler or dynamic scheduler supported with response-time analysis.

- *Execution stack*. These two components are in charge of orchestrating the parallel execution of the application in a time-predictable manner, based on the DAG.

1.5 Summary

Providing high performance while meeting predictability requirements of real-time applications is a challenging task, which requires new techniques and tools at most if not all levels of the design flow and execution stack. This book presents the work which was done within the P-SOCRATES project to address these challenges, presenting solutions for deriving control- and data-flow graph of OpenMP parallel programs using the tasking model, algorithms for mapping and scheduling the OpenMP tasks into many-core platforms, and methods to perform both timing and schedulability analysis. The book also describes solutions for the runtime execution stack for real-time parallel computation, both at the level of the OpenMP runtime, as well as within real-time operating systems.

References

[1] Magid, Y., Adi, A., Barnea, M., Botzer, D., Rabinovich, E., "Application generation framework for real-time complex event processing," *32nd Annual IEEE International Computer Software and Applications (COMPSAC)*, 2008.

[2] Anicic, D., Rudolph, S., Fodor, P., Stojanovic, N., "Stream reasoning and complex event processing in ETALIS," *Semantic Web 1*, 2009, IOS Press, pp. 1–5.

[3] Luckham, D. C., "*Event Processing for Business: Organizing the Real-Time Enterprise,*" John Wiley and Sons, 2011.

[4] Palmer, M., "*Real-Time Big Data and the 11 Principles of Modern Surveillance Systems*," http://streambase.typepad.com/streambase_stream_process/2011/07/in-his-tabbforum-article-dave-tolladay-eloquently-argues-that-real-time-surveillance-is-crucial-in-todays-high-frequency-t.html, last accessed February 2018.

[5] Twentyman, J., "*Sensory Perception,*" http://www.information-age.com/technology/information-management/1248733/sensory-perception, last accessed February 2018.

[6] Klein, R., Xie, J., and Usov, A., "Complex events and actions to control cyber-physical systems." In *Proceedings of the 5th ACM International Conference on Distributed Event-Based System (DEBS)*, 2011.

[7] Shapiro, M., "Supersymmetry, extra dimensions and the origin of mass: exploring the nature of the universe using petaScale data analysis," *Google TechTalk*, June 18, 2007.

[8] "*NTT DATA: Staying Ahead of the IT Services Curve With Real-Time Analytics,*" https://www.sap.com/sea/documents/2012/10/66e7c78d-357c-0010-82c7-eda71af511fa.html, last accessed February 2018.

[9] "*SAP Enters Complex-event Processing Market,*" http://www.cio.com.au/article/377688/sap_enters_complex-event_processing_market/, last accessed February 2018.

[10] Tieman, R., "Algo trading: the dog that bit its master", *Financial Times*, March 2008.

[11] Karim, L., Boulmakoul, A., Lbath, A., "Near real-time big data analytics for NFC-enabled logistics trajectories," *2016 3rd International Conference on Logistics Operations Management (GOL)*, Fez, 2016, pp. 1–7.

[12] Ungerer, T., et. al. "MERASA: Multi-core execution of hard real-time applications supporting analysability," In *the IEEE Micro 2010, Special Issue on European Multicore Processing Projects*, Vol. 30, No. 5, October 2010.

[13] Lundqvist, T., Stenstrom, P., "Timing anomalies in dynamically scheduled microprocessors." In *IEEE Real-Time Systems Symposium*, 1999.

[14] "*OpenCL (Open Computing Language)*", http://www.khronos.org/opencl, last accessed February 2018.

[15] *NVIDIA*, https://developer.nvidia.com/cuda-zone, last accessed February 2018.

2

Manycore Platforms

Andrea Marongiu[1], Vincent Nélis[2] and Patrick Meumeu Yomsi[2]

[1]Swiss Federal Institute of Technology in Zürich (ETHZ), Switzerland; and University of Bologna, Italy
[2]CISTER Research Centre, Polytechnic Institute of Porto, Portugal

This chapter surveys state-of-the-art manycore platforms. It discusses the historical evolution of computing platforms over the past decades and the technical hurdles that led to the manycore revolution, then presents in details several manycore platforms, outlining (i) the key architectural traits that enable scalability to several tens or hundreds of processing cores and (ii) the shared resources that are responsible for unpredictable timing.

2.1 Introduction

Starting from the early 2000s, general-purpose processor manufacturers adopted the chip multiprocessor (CMP) design paradigm [1] to overcome technological "walls."

Single-core processor designs hit the **power wall** around 2004, when the consolidated strategy of scaling down the gate size of integrated circuits – reducing the supply voltage and increasing the clock frequency – became unfeasible because of excessive power consumption and expensive packaging and cooling solutions [2]. The CMP phisolophy replaces a single, very fast core with multiple cores that cooperate to achieve equivalent performance, but each operating at a lower clock frequency and thus consuming less power.

Over the past 20 years, processor performance has increased at a faster rate than the memory performance [3], which created a gap that is commonly referred to as the **memory wall**. Historically, sophisticated multi-level cache hierarchies have been built to implement main memory access latency hiding techniques. As CMPs use lower clock frequencies, the processor–memory

15

gap grows at a slower rate, compared to traditional single-core systems. Globally, the traditional latency hiding problem is turned into an increased bandwidth demand, which is easier to address, as the DRAM bandwidth scales much better than its access latency [4].

Single-core designs have traditionally been concerned with the development of techniques to efficiently extract instruction-level parallelism (ILP). However, increasing ILP performance beyond what is achieved today with state-of-the-art techniques has become very difficult [5], which is referred to as the **ILP wall**. CMPs solve the problem by shifting the focus to thread-level parallelism (TLP), which is exposed at the parallel programming model level, rather than designing sophisticated hardware to transparently extract ILP from instruction streams.

Finally, the **complexity** wall refers to the difficulties encountered by single-core chip manifacturers in designing and verifying increasingly sophisticated out-of-order processors. In the CMP design paradigm, a much simpler processor core is designed once and replicated to scale to the multicore system core count. Design reuse and simplified core complexity obviously significantly reduce the system design and verification.

The trend towards integrating an increasing number of cores in a single chip has continued all over the past decade, which has progressively paved the way for the introduction of manycore systems, i.e., CMPs containing a high number of cores (tens to hundreds). Interestingly, the same type of "revolution" has taken place virtually in every domain, from the high-performance computing (HPC) to the embedded systems (ES). Driven by converging needs for high performance requirements, energy efficiency, and flexibility, the most representative commercial platforms from both domains nowadays feature very similar architectural traits. In particular, core *clusterization* is the key design paradigm adopted in all these products. A hierarchical processor organization is always employed, where simple processing units are grouped into small-medium sized subsystems (the *clusters*) and share high-performance local interconnection and memory. Scaling to larger system sizes is enabled by replicating clusters and interconnecting them with a scalable medium like a network-on-chip (NoC).

In the following, we briefly present several manycore platforms, both from the HPC and the ES domains. We discuss the Kalray MPPA-256 at last, and in greater detail, as this is the platform for which the development of the software techniques and the experimental evaluation presented throughout the rest of the book have been conducted.

2.2 Manycore Architectures

2.2.1 Xeon Phi

Xeon Phi are a series of x86 manycore processors by Intel and meant to accelerate the highly parallel workloads of the HPC world. As such, they are employed in supercomputers, servers, and high-end workstations. The Xeon Phi family of products has its roots in the *Larrabee* microarchitecture project – an attempt to create a manycore accelerator meant as a GPU as well as for general-purpose computing – and has recently seen the launch of the Knights Landing (KNL) chip on the marketplace.

Figure 2.1a shows the high-level block diagram of the KNL CPU. It comprises 38 physical *tiles*, of which at most 36 are active (the remaining two tiles are for yield recovery). The structure of a *tile* is shown in Figure 2.1b. Each *tile* comprises two cores, two vector processing units (VPUs) per core, and a 1-Mbyte level-2 (L2) cache that is shared between the two cores.

The core is derived from the Intel Atom (based on the Silvermont microarchitecture [6]), but leverages a new two-wide, out-of-order core which includes heavy modifications to incorporate features necessary for HPC workloads [e.g., four threads per core, deeper out-of-order buffers, higher cache bandwidth, new instructions, better reliability, larger translation look-aside buffers (TLBs), and larger caches]. In addition, the new Advanced

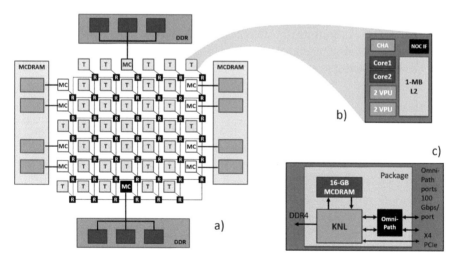

Figure 2.1   Knights Landing (KNL) block diagram: (a) the CPU, (b) an example tile, and (c) KNL with Omni-Path Fabric integrated on the CPU package.

Vector Extensions instruction set, AVX-512, provides 512-bit-wide vector instructions and more vector registers.

At the top level, a 2D, cache-coherent mesh NoC connects the tiles, memory controllers, I/O controllers, and other agents on the chip. The mesh supports the MESIF (modified, exclusive, shared, invalid, forward) protocol, which employs a distributed tag directory to keep the L2 caches in all tiles coherent with each other. Each tile contains a caching/home agent that holds a portion of the distributed tag directory and also serves as a connection point between the tile and the mesh.

Knights Landing features two types of memory: (i) multichannel DRAM (MCDRAM) and (ii) double data rate (DDR) memory. MCDRAM is organized as eight devices – each featuring 2-Gbyte high-bandwidth banks – integrated on-package and connected to the KNL die via a proprietary on-package I/O. The DDR4 is organized as six channels running at up to 2,400 MHz, with three channels on each of two memory controllers.

The two types of memory are presented to users in three memory modes: cache mode, in which MCDRAM is a cache for DDR; flat mode, in which MCDRAM is treated like standard memory in the same address space as DDR; and hybrid mode, in which a portion of MCDRAM is cache and the remainder is flat. KNL supports a total of 36 lanes of PCI express (PCIe) Gen3 for I/O, split into two x16 lanes and one x4 lane. Moreover, it integrates the Intel Omni-Path Fabric on-package (see Figure 2.1c), which provides two 100-Gbits-per-second ports out of the package.

The typical power (thermal design power) for KNL (including MCDRAM memory) when running a computationally intensive workload is 215 W without the fabric and 230 W with the fabric.

2.2.2 Pezy SC

PEZY-SC (PEZY Super Computer) [7] is the second generation manycore microprocessor developed by PEZY in 2014, and is widely used as an accelerator for HPC workloads. Compared to the original PEZY-1, the chip contains exactly twice as many cores and incorporates a large amount of cache including 8 MB of L3$. Operating at 733 MHz, the processor is said to have peak performance of 3.0 TFLOPS (single-precision) and 1.5 TFLOPS (double-precision). PEZY-SC was designed using 580 million gates and manufactured on TSMC's 28HPC+ (28 nm process).

In June 2015, PEZY-SC-based supercomputers took all top three spots on the Green500 listing as the three most efficient supercomputers:

1. **Shoubu**: 1,181,952 cores, 50.3 kW, 605.624 TFlop/s Linpack Rmax;
2. **Suiren Blue**: 262,656 cores, 40.86 kW, 247.752 TFlop/s Linpack Rmax;
3. **Suiren**: 328,480 cores, 48.90 kW, 271.782 TFlop/s Linpack Rmax.

PEZY-SC contains two ARM926 cores (ARMv5TEJ) along with 1024 simpler RISC cores supporting 8-way SMT for a total of 8,192 threads, as shown in Figure 2.2. The organization of the accelerator cores in PEZY-SC heavily uses clusterization and hierarchy. At the top level, the microprocessor is made of four blocks called "*prefectures.*" Within a *prefecture*, 16 smaller blocks called "*cities*" share 2 MB of L3$. Each *city* is composed of 64 KB of shared L2$, a number of special function units and four smaller blocks called "*villages.*" Inside a *village* there are four execution units and every two such execution units share 2 KB of L1D$.

The chip has a peak power dissipation of 100 W with a typical power consumption of 70 W which consists of 10 W leakage + 60 W dynamic.

2.2.3 NVIDIA Tegra X1

The NVIDIA Tegra X1 [8] is a hybrid System on Module (SoM) featured in the NVIDIA Jetson Development boards. As a mobile processor, the Tegra X1 is meant for the high-end ES markets, and is the first system to feature a

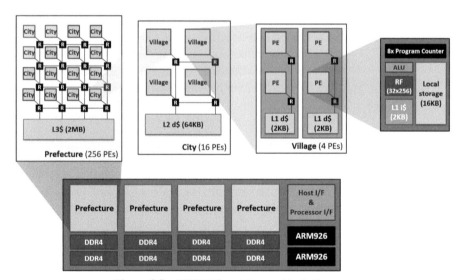

Figure 2.2 PEZY-SC architecture block diagram.

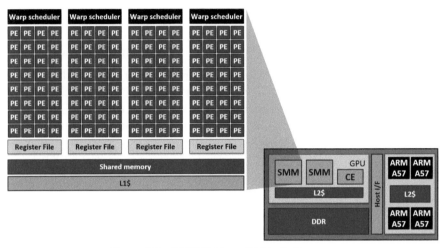

Figure 2.3   NVIDIA Tegra X1 block diagram.

chip powerful enough to sustain the visual computing load for autonomous and assisted driving applications.

As shown in Figure 2.3, the X1 CPU complex consists of a big LIT-TLE architecture, featuring quad-core 1.9 GHz ARM Cortex-A57 processor (48 KB I-cache + 32 kB D-cache L1 per core, 2 MB L2 cache common to all cores), plus quad-core ARM Cortex A53 processor. A single CPU core can utilize the maximum bandwidth available for the whole CPU complex, which amounts to almost 4.5 GB/s for sequential read operations.

The iGPU is a second-generation Maxwell "GM20b" architecture, with 256 CUDA cores grouped in two Streaming Multi-processors (SMs) (the "*clusters*") sharing a 256 KB L2 (last-level) cache. The compute pipeline of an NVIDIA GPU includes engines responsible for computations (Execution Engine, EE) and engines responsible for high bandwidth memory transfers (Copy Engine, CE). The EE and CE can access central memory with a maximum bandwidth close to 20 GB/s, which can saturate the whole DRAM bandwidth. Indeed, the system DRAM consists of 4 GB of LPDDR4 64 bit SDRAM working at (maximum) 1.6 GHz, reaching a peak ideal bandwidth of 25.6 GB/s.

Despite the high performance capabilities of the SoC (peak performance 1 TFlops single precision), the Tegra X1 features a very contained power envelope, drawing 6–15 W.

2.2.4 Tilera Tile

The *Tile* architecture has its roots in the RAW research processor developed at MIT [9] and later commercialized by Tilera, a start-up founded by the original research group. Chips from the second generation are expected to scale up to 100 cores based on the MIPS ISA and running at 1.5 GHz.

The *Tile* architecture is among the first examples of a cluster-based many-core, featuring *ad-hoc* on-chip interconnect and cache architecture. The architectural template is shown in Figure 2.4. The chip is architected as a 2D array of *tiles* (the *clusters*), interconnected via a mesh-based NoC. Each tile contains a single processor core, with local L1 (64 KB) and a portion (256 KB) of the distributed L2 cache. Overall, the L2 cache segments behave as a non-uniformly addressed cache (NUCA), using a directory-based coherence mechanism and the concept of *home tile* (the tile that holds the master copy) for cached data. The NUCA design makes cache access latency variable according to the distance between tiles, but enables an efficient (space- and power-wise) logical view to the programmer: a large on-chip cache to which all cores are connected. Each tile also features an interconnect switch that connects it to the neighboring tiles, which allows for a simplified interconnect design (essentially, a switched network with very short wires connecting neighboring tiles linked through the tile-local switch).

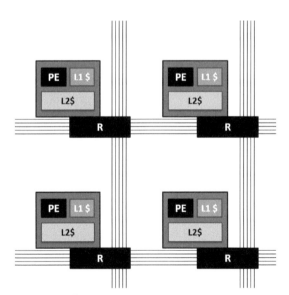

Figure 2.4   Tilera *Tile* architectural template.

The NoC – called *iMesh* by Tilera – actually consists of five different networks, used for various purposes:

- Application process communication (UDN),
- I/O communication (IDN),
- Memory communication (MDN),
- Cache coherency (TDN),
- Static, channelized communication (STN).

The latency of the data transfers on the network is 1–2 cycles/tile, depending on whether there's a direction change or not at the tile. The *TileDirect* technology allows data received over the external interfaces to be placed directly into the tile-local memory, thus bypassing the external DDR memory and reducing memory traffic.

The power budget of the *Tile* processors is under 60 W.

2.2.5 STMicroelectronics STHORM

STHORM is a heterogeneous, manycore-based system from STMicroelectronics [10], with an operating frequency ranging up to 600 MHz.

The STHORM architecture is organized as a fabric of multi-core *clusters*, as shown in Figure 2.5. Each cluster contains 16 STxP70 *Processing*

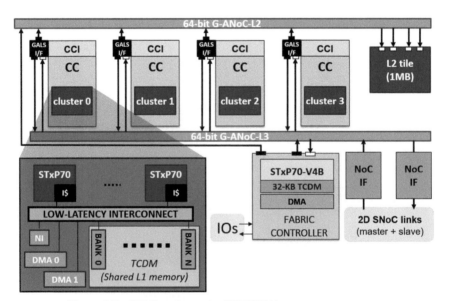

Figure 2.5   STMicroelectronics STHORM heterogeneous system.

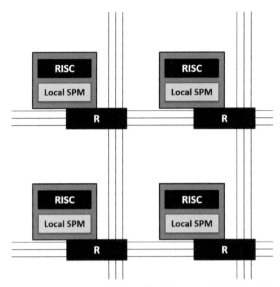

Figure 2.6 Block diagram of the Epiphany-V chip from Adapteva.

Elements (PEs), each of which has a 32-bit dual-issue RISC processor. PEs communicate through a shared multi-ported, multi-bank, tightly-coupled data memory (TCDM, a scratchpad memory). Additionally, STHORM clusters feature an additional core called the *cluster controller* (CC) and meant, as the name suggests, for the execution of control code local to the cluster operation. Globally, four *clusters* plus a *fabric controller* (FC) core – responsible for global coordination of the clusters – are interconnected via two asynchronous networks-on-chip (ANoC). The first ANoC is used for accessing a multi-banked, multiported L2 memory, shared among the four clusters. The second ANoC is used for inter-cluster communication via L1 TCDMs (i.e., remote clusters' TCDMs can be accessed by every core in the system) and to access the offchip main memory (L3 DRAM).

STHORM delivers up to 80 GOps (single-precision floating point) with only 2W power consumption.

2.2.6 Epiphany-V

The Epiphany-V chip from Adapteva [11] is based on a 1024-core processor in 16 nm FinFet technology. The chip contains an array of 1024 64-bit RISC processors, 64 MB of on-chip SRAM, three 136-bit wide mesh Networks-On-Chip, and 1,024 programmable IO pins.

Similar to the Tilera *Tile* architecture, the Epiphany architecture is a distributed shared memory architecture composed of an array of RISC processors communicating via a low-latency, mesh-based NoC, as shown in Figure 2.6. Each cluster (or *node*) in the 2D array features a single, complete RISC processor capable of independently running an operating system [according to the multiple-instruction, multiple-data (MIMD) paradigm]. The distributed shared memory model of the Epiphany-V chip relies on a cacheless design, in which all scratchpad memory blocks are readable and writable by all processors in the system (similar to the STHORM chip).

The Epiphany-V chip can deliver two teraflops of performance (single-precision floating point) in a 2W power envelope.

2.2.7 TI Keystone II

The Texas Instrument Keystone II [12], is a heterogeneous SoC featuring a quad-core ARM Cortex-A15 and an accelerator cluster comprising eight C66x VLIW DSPs. The chip is designed for special-purpose industrial tasks, such as networking, automotive, and low-power server applications. The 66AK2H12 SoC, depicted in Figure 2.7, is the top-performance Texas Instrument Keystone II device architecture.

Each DSP in the accelerator cluster is a VLIW core, capable of fetching up to eight instructions per cycle and running at up to 1.2 GHz. Locally,

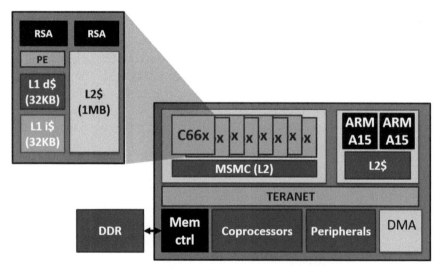

Figure 2.7   Texas Instrument Keystone II heterogeneous system.

a DSP is equipped with 32 KB L1 D-cache and L1 I-cache, plus 1024 KB L2 unified cache. Altogether, the DSPs in the accelerator cluster deliver 160 single-precision GOps.

On the ARM side, there are 32 KB of L1 D-cache and 32 KB of L1 I-cache per core, plus a coherent 4 MB L2 cache.

The computational power of such architecture, at a power budget of up to 14 W, makes it a low-power solution for microserver-class applications. The Keystone II processor has been used in several cloud-computing/microserver settings [13–15].

2.2.8 Kalray MPPA-256

The Kalray MPPA-256 processor of the MPPA (Multi-Purpose Processor Array) MANYCORE family has been developed by the company KALRAY. It is a single-chip programmable manycore processor manufactured in 28 nm CMOS technology that targets low-to-medium volume professional applications, where low energy per operation and time predictability are the primary requirements [16]. It concentrates a great potential and is very promising for high-performance parallel computing. With an operating frequency of 400 MHz and a typical power consumption of 5 W, the processor can perform up to 700 GOPS and 230 GFLOPS. The processor integrates a total of 288 identical Very Long Instruction Word (VLIW) cores including 256 user cores referred to as processing engines (PEs) and dedicated to the execution of the user applications and 32 system cores referred to as Resource Manager (RM) and dedicated to the management of the software and processing resources. The cores are organized in 16 compute clusters and four I/O subsystems to control all the I/O devices. In Figure 2.8, the 16 inner nodes (labeled CC)

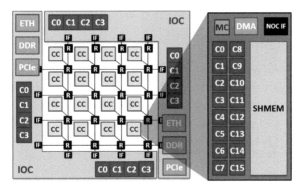

Figure 2.8 High-level view of the Kalray MPPA-256 processor.

correspond to the 16 compute clusters holding 17 cores each: 16 PEs and 1 RM. Then, there are four I/O subsystems located at the periphery of the chip, each holding four RMs. Each compute cluster and I/O subsystem owns a private address space, while communication and synchronization between them is ensured by the data and control NoC depicted in Figure 2.8. The MPPA-256 processor is also fitted with a variety of I/O controllers, in particular DDR, PCI, Ethernet, Interlaken, and GPIO.

2.2.8.1 The I/O subsystem

The four I/O subsystems (also denoted as IOS) are referenced as the North, South, East, and West IOS. They are responsible for all communications with elements outside the MPPA-256 processor, including the host workstation if the MPPA is used as an accelerator.

Each IOS contains four RMs in a symmetric multiprocessing configuration. These four RMs are connected to a shared, 16-bank parallel memory of 512 KB, they have their own private instruction cache of 32 KB (8-way, set-associative) and share a data cache of 128 KB (also 8-way, set-associative), which ensures data coherency between the cores.

The four IOS are dedicated to PCIe, Ethernet, Interlaken, and other I/O devices. Each one runs either a rich OS such as Linux or an RTOS that supports the MPPA I/O device drivers. They integrate controllers for an 8-lane Gen3 PCIe for a total peak throughput of 16 GB/s full duplex, Ethernet links ranging from 10 MB/s to 40 GB/s for a total aggregate throughput of 80 GB/s, the Interlaken link providing a way to extend the NoC across MPPA-256 chips and other I/O devices in various configurations like UARTs, I2C, SPI, pulse width modulator (PWM), or general purpose IOs (GPIOs). More precisely, the East and West IOS are connected to a quad 10 GB/s Ethernet controller, while the North and South IOS are connected to an 8-lane PCIe controller and to a DDR interface for access to up to 64 GB of external DDR3-1600.

2.2.8.2 The Network-on-Chip (NoC)

The NoC holds a key role in the average performance of manycore architectures, especially when different clusters need to exchange messages. In the Kalray MPPA-256 processor, the 16 compute clusters and the four I/O subsystems are connected by two explicitly addressed NoC with bi-directional links providing a full duplex bandwidth up to 3.2 GB/s between two adjacent nodes:

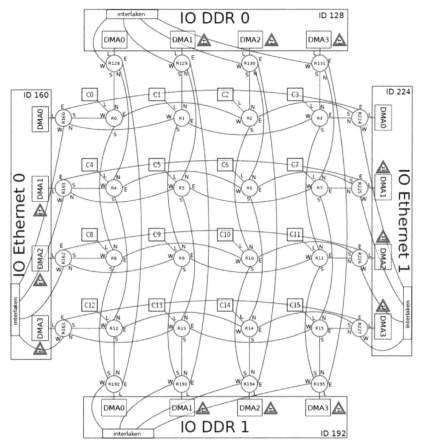

Figure 2.9   MPPA-256 NoC architecture.

- The data NoC (D-NoC). This NoC is optimized for bulk data transfers;
- The control NoC (C-NoC). This NoC is optimized for small messages at low latency.

The two NoCs are identical with respect to the nodes, the 2D-wrapped-around torus topology, shown in Figure 2.9, and the wormhole route encoding. They differ at their device interfaces, by the amount of packet buffering in routers, and by the flow regulation at the source available on the D-NoC. NoC traffic through a router does not interfere with the memory buses of the underlying I/O subsystem or compute cluster, unless that router is the destination node. Besides, the D-NoC implements a quality-of-service (QoS) mechanism, thus guaranteeing predictable latencies for all data transfers.

2.2.8.3 The Host-to-IOS communication protocol

The special hierarchy among the cores in the MPPA-256 processor helps to better divide the workload to be executed on the PEs. When the MPPA-256 is used as an accelerator, tasks are sent to the MPPA-256 processor from a Host workstation. The communication with the MPPA-256 can thus be performed in a couple of steps which can be referred to as Host-to-IOS, IOS-to-Clusters and finally Cluster-to-Cluster communication protocols. The MPPA-256 processor communicates with the Host workstation through I/O subsystems. The chip is connected to the host CPU by a PCIe interface and two connectors – **Buffer** and **MQueue** – are available for making this link. The RM core that accommodates the task upon the I/O subsystem is referred to as **Master** (see Figure 2.10). The processor then executes the received task (referred to as **Master task**) as detailed in Section 4.3.1 and at the end of the execution process, it writes the output data in a 4 GB DDR3 RAM memory, which is connected to an I/O subsystem and can be accessed by the host CPU.

2.2.8.4 Internal architecture of the compute clusters

The compute cluster (Figure 2.11) is the basic processing unit of the MPPA architecture. Each cluster contains 17 Kalray-1 VLIW cores, including 16 PE cores dedicated to the execution of the user applications and one RM core. Among other responsibilities, the RM is in charge of mapping and scheduling the threads on the PEs and managing the communications between the clusters and between the clusters and the main memory. The 16 PEs and the RM

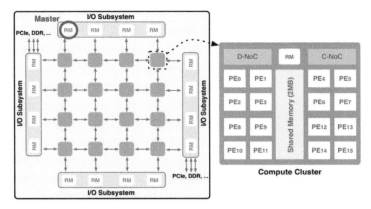

Figure 2.10   A master task runs on an RM of an I/O subsystem.

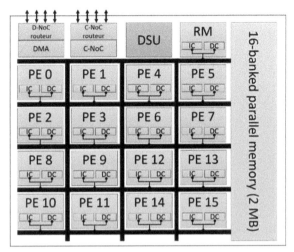

Figure 2.11 Internal architecture of a compute cluster.

are connected to a shared memory of 2 MB. A direct memory access (DMA) engine is responsible for transferring data between the shared memory and the NoC or within the shared memory. The DMA engine supports multi-dimensional data transfers and sustains a total throughput of 3.2 GB/s in full duplex. The Debug and System Unit (DSU) supports the compute cluster debug and diagnostics capabilities. Each DSU is connected to the outside world by a JTAG (IEEE 1149.1) chain. The DSU also contains a system trace IP that is used by lightly instrumented code to push up to 1.6 GB/s of trace data to an external acquisition device. This trace data gives almost non-intrusive insight on the behaviour of the application.

2.2.8.5 The shared memory

The shared memory (SMEM) in each compute cluster (yellow box in Figure 2.11) comprises 16-banked independent memory of 16,384 x 64-bit words = 128 kB per bank, with a total capacity of 16 x 128 kB = 2 MB, with error code correction (ECC) on 64-bit words. This memory space is shared between the 17 VLIW cores in the cluster and delivers an aggregate bandwidth of 38.4 GB/s.

The 16 memory banks are arranged in two sides of eight banks, the left side and the right side. The connections between the memory bus masters are replicated in order to provide independent access to the two sides. There are two ways of mapping a physical address to a specific side and bank.

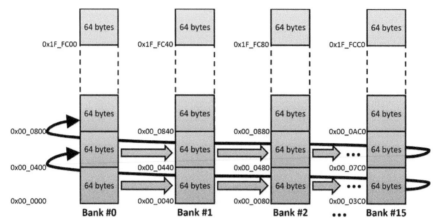

Figure 2.12 Memory accesses distributed across memory banks (interleaved).

Option 1 (Interleaving address mapping) – In the address space, bits 6–9 of the byte address select the memory bank, so sequential addresses move from one bank to another every 64 bytes (every 8 x 64-bit words), as depicted in Figure 2.12. This address-mapping scheme is effective at distributing the requests of cores across memory banks, while ensuring that each cache refill request involves only one memory bank and benefits from a burst access mode. Furthermore, this address scheme also allows the "simultaneous" access (respecting the activation time) of those memory banks in which the cache line is stored. As the side selection depends on the sixth bit of the byte address, the bank selection by sequential addresses alternates between the left side and the right side every 64 bytes.

Option 2 (Contiguous address mapping) – It is possible to disable the memory address shuffling, in which case each bank has a sequential address space covering one bank of 128 KB as depicted in Figure 2.13. The high-order bit of the address selects the side (i.e., the right side covers addresses from 0 to 1 MB and the left side covers addresses above 1 MB). When zero interference between cores is needed, cores within a given pair must use a different side.

2.3 Summary

Back in the early days of the new millennium, multicore processors allowed computer designers to overcome several technological *walls* that traditional single-core design methodologies were no longer capable of addressing. This

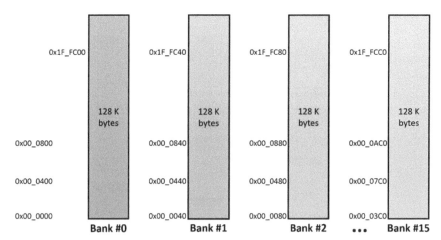

Figure 2.13 Memory accesses targeting a same memory bank (contiguous).

design paradigm is to date the standard, with an ever-increasing number of processing cores integrated on the same chip. While manycore processors enabled over the past 15 years the seamless continuation of compute performance scalability for general-purpose and scientific workloads, real-time systems have not been able to embrace this technology so far, due to the lack of predictability in execution time implied by hardware resource sharing. This chapter has surveyed several state-of-the-art manycore processors, highlighting the architectural features (i) that enable processor integration scalability and (ii) those that are shared among several processor and that are mostly responsible for the unpredictable execution.

References

[1] Olukotun, K., Nayfeh, B. A., Hammond, L., Wilson, K., and Chang, K., The case for a single-chip multiprocessor. *SIGOPS Oper. Syst. Rev.*, 30, 2–11, 1996.
[2] Fuller, S. H., and Millett, L. I., Computing performance: Game over or next level? *Computer*, 44, 31–38, 2011.
[3] Hennessy, J. L., and Patterson, D. A., *Computer Architecture: A Quantitative Approach*. Elsevier, 2011.
[4] Patterson, D. A., Latency lags bandwith. *Commun. ACM*, 47, 71–75, 2004.

[5] Agarwal, V., Hrishikesh, M. S., Keckler, S. W., and Burger, D., "Clock rate versus ipc: the end of the road for conventional microarchitectures." In *Proceedings of 27th International Symposium on Computer Architecture (IEEE Cat. No.RS00201)*, pages 248–259, 2000.

[6] Naffziger, S., and Sohi, G., Hot chips 26. *IEEE Micro*. 35, 4–5, 2015.

[7] Tabuchi, A., Kimura, Y., Torii, S., Matsufuru, H., Ishikawa, T., Boku, T., and Sato, M., *Design and Preliminary Evaluation of Omni OpenACC Compiler for Massive MIMD Processor PEZY-SC*, pp. 293–305, Springer International Publishing, Cham, 2016.

[8] NVIDIA SRL. *Whitepaper: NVIDIA Tegra X1 – NVIDIA's New Mobile Superchip.* https://docs.nvidia.com/cuda/cuda-c-programming-guide/index.html, accessed November 07, 2011.

[9] Taylor, M. B., Kim, J., Miller, J., Wentzlaff, D., Ghodrat, F., Greenwald, B., et al. The raw microprocessor: a computational fabric for software circuits and general-purpose programs. *IEEE Micro*, 22, 25–35, 2002.

[10] Melpignano, D., Benini, L., Flamand, E., Jego, B., Lepley, T., Haugou, G., Clermidy, F., and Dutoit, D., "Platform 2012, a many-core computing accelerator for embedded socs: Performance evaluation of visual analytics applications." In *Proceedings of the 49th Annual Design Automation Conference*, DAC '12, pp. 1137–1142, New York, NY, USA, 2012.

[11] Olofsson, A., Epiphany-v: A 1024 processor 64-bit RISC system-on-chip. *CoRR*, abs/1610.01832, 2016.

[12] Stotzer, E., Jayaraj, A., Ali, M., Friedmann, A., Mitra, G., Rendell, A. P., and Lintault, I., *OpenMP on the Low-Power TI Keystone II ARM/DSP System-on-Chip*, pp. 114–127, Springer Berlin Heidelberg, Berlin, Heidelberg, 2013.

[13] Verma, A., and Flanagan, T., A Better Way to Cloud. *Texas Instruments white paper*, 2012.

[14] Hewlett-Packard Development Company L.P. *HP ProLiant m800 Server Cartridge*.

[15] nCore HPC LLC. *Brown Dwarf Y-Class Supercomputer*.

[16] Amdahl, G. M., "Validity of the single processor approach to achieving large scale computing capabilities." In *Proceedings of the April 18–20, 1967, Spring Joint Computer Conference, AFIPS '67* (Spring), pp. 483–485, New York, NY, USA, 1967.

3

Predictable Parallel Programming with OpenMP

Maria A. Serrano[1], Sara Royuela[1], Andrea Marongiu[2] and Eduardo Quiñones[1]

[1]Barcelona Supercomputing Center (BSC), Spain
[2]Swiss Federal Institute of Technology in Zürich (ETHZ), Switzerland; and University of Bologna, Italy

This chapter motivates the use of the OpenMP (Open Multi-Processing) parallel programming model to develop future critical real-time embedded systems, and analyzes the time-predictable properties of the OpenMP tasking model. Moreover, this chapter presents the set of compiler techniques needed to extract the timing information of an OpenMP program in the form of an *OpenMP Direct Acyclic Graph* or OpenMP-DAG.

3.1 Introduction

Parallel programming models are key to increase the productivity of parallel software from three different angles:

1. From a *programmability* angle, parallel programming models provide developers with the abstraction level required to program parallel applications while hiding processor complexities.
2. From a *portability* angle, platform-independent parallel programming models allow executing the same parallel source code in different parallel platforms.
3. From a *performance* angle, different levels of abstraction allow for a fine-tuned parallelism, i.e., users may either squeeze the capabilities of a specific architecture using the language capabilities, or rely on runtime mechanisms to dynamically exploit parallelism.

Hence, parallel programming models are of paramount importance to exploit the massive computation capabilities of state-of-the-art and future parallel and heterogeneous processor architectures. Several approaches coexist with such a goal, and these can be grouped as follows [1]:

- *Hardware-centric models* aim to replace the native platform programming with higher-level, user-friendly solutions, e.g., Intel® TBB [2] and NVIDIA® CUDA [3]. These models focus on tuning an application to match a chosen platform, which makes their use neither a scalable nor a portable solution.
- *Application-centric models* deal with the application parallelization from design to implementation, e.g., OpenCL [4]. Although portable, these models may require a full rewriting process to accomplish productivity.
- *Parallelism-centric models* allow users to express typical parallelism constructs in a simple and effective way, and at various levels of abstraction, e.g., POSIX threads [6] and OpenMP [7]. This approach allows flexibility and expressiveness, while decoupling design from implementation.

Considering the vast amount of parallel programming models available, there is a noticeable need to unify programming models to exploit the performance benefits of parallel and heterogeneous architectures [9]. In that sense, OpenMP has proved many advantages over its competitors to enhance productivity. The next sections introduce the main characteristics of the most relevant programming models, and conclude with an analysis of the main benefits of OpenMP.

3.1.1 Introduction to Parallel Programming Models

The multitude of parallel programming models currently existing makes it difficult to choose the language that better fits the needs of each particular case. Table 3.1 introduces the main characteristics of the most relevant programming models in critical embedded systems. The features considered are the following: performance (based on throughput, bandwidth, and other metrics), portability (based on how straight-forward it is to migrate to different environments), heterogeneity (based on the support for cross-platform applications), parallelism (based on the support provided for data-based and task-based parallelism), programmability (based on how easy it is for programmers to get the best results), and flexibility (based on the features for parallelizing offered in the language).

Table 3.1   Parallel programming models comparison

	Pthreads	OpenCL	CUDA	Cilk Plus	TBB	OpenMP
Performance	✓	✓	✓✓	✓✓	✓	✓
Portability	✓	✓	×	×	×	✓✓
Heterogeneity	×	✓	✓	×	✓	✓✓
Parallelism	data/task	data/task	data	data/task	task	data/task
Programmability	×	×	×	✓✓	✓	✓
Flexibility	✓	✓	×	×	✓	✓✓

3.1.1.1 POSIX threads

POSIX threads (Portable Operating System Interface for UNIX threads), usually referred to as Pthreads, is a standard C language programming interface for UNIX systems. The language provides efficient light-weight mechanisms for thread management and synchronization, including mutual exclusion and barriers.

In a context where hardware vendors used to implement their own proprietary versions of threads, Pthreads arose with the aim of enhancing the portability of threaded applications that reside on shared memory platforms. However, Pthreads results in very poor programmability, due to the low-level threading model provided by the standard, that leaves most of the implementation details to the programmer (e.g., work-load partitioning, worker management, communication, synchronization, and task mapping). Overall, the task of developing applications with Pthreads is very hard.

3.1.1.2 OpenCL™

OpenCL™ (Open Computing Language) is an open low-level application programming interface (API) for cross-platform parallel computing that runs on heterogeneous systems including multicore and manycore CPUs, GPUs, DSPs, and FPGAs. There are two different actors in an OpenCL system: the host and the devices. The language specifies a programming language based on C99 used to control the host, and a standard interface for parallel computing, which exploits task-based and data-based parallelism, used to control the devices.

OpenCL can run in a large variety of devices, which makes portability its most valuable characteristic. However, the use of vendor-specific features may prevent this portability, and codes are not guaranteed to be optimal due to the important differences between devices. Furthermore, the language has an important drawback: it is significantly difficult to learn, affecting the programmability.

3.1.1.3 NVIDIA® CUDA

NVIDIA® CUDA is a parallel computing platform and API for exploiting CUDA-enabled GPUs for general-purpose processing. The platform provides a layer that gives direct access to the GPU's instruction set, and is accessible through CUDA-accelerated libraries, compiler directives (such as OpenACC [10]), and extensions to industry-standard programming languages (such as C and C++).

The language provides dramatic increases of performance when exploiting parallelism in GPGPUs. However, its use is limited to CUDA-enabled GPUs, which are produced only by NVIDIA®. Furthermore, tuning applications with CUDA may be hard because it requires rewriting all the offloaded kernels and knowing the specifics of each platform to get the best results.

3.1.1.4 Intel® Cilk™ Plus

Intel® Cilk Plus [11] is an extension to C/C++ based on Cilk++ [12] that has become popular because of its simplicity and high level of abstraction. The language provides support for both data and task parallelism, and provides a framework that optimizes load balance, implementing a work-stealing mechanism to execute tasks [13].

The language provides a simple yet efficient platform for implementing parallelism. Nonetheless, portability is very limited because only Intel® and GCC implement support for the language extensions defined by Cilk Plus. Furthermore, the possibilities available with this language are limited to tasks (_cilk_spawn, _cilk_sync), loops (_cilk_for), and reductions (*reducers*).

3.1.1.5 Intel® TBB

Intel® TBB is an object-oriented C++ template library for implementing task-based parallelism. The language offers constructs for parallel loops, reductions, scans, and pipeline parallelism. The framework provided has two key components: (1) compilers, which optimize the language templates enabling a low-overhead form of polymorphism, and (2) runtimes, which keep temporal locality by implementing a queue of tasks for each worker, and balance workload across available cores by implementing a work-stealing policy.

TBB offers a high level of abstraction in front of complicated low-level APIs. However, adapting the code to fit the library templates can be arduous. Furthermore, portability is limited, although the last releases support Visual C++, Intel® C++ compiler, and the GNU compiler collection.

3.1.1.6 OpenMP

OpenMP, the de-facto standard parallel programming model for shared memory architectures in the high-performance computing (HPC) domain, is increasingly adopted also in embedded systems. The language was originally focused on a thread-centric model to exploit massive data-parallelism and loop intensive applications. However, the latest specifications of OpenMP have evolved to a task-centric model that enables very sophisticated types of fine-grained and irregular parallelism, and also include a host-centric accelerator model that enables an efficient exploitation of heterogeneous systems. As a matter of fact, OpenMP is supported in the SDK of many of the state-of-the-art parallel and heterogeneous embedded processor architectures, e.g., Kalray MPPA [14], and TI Keystone II [16].

Different evaluations demonstrate that OpenMP delivers tantamount performance and efficiency compared to highly tunable models such as TBB [17], CUDA [18] and OpenCL [19]. Moreover, OpenMP has different advantages over low-level libraries such as Pthreads: on one hand, it offers robustness without sacrificing performance [21] and, on the other hand, OpenMP does not lock the software to a specific number of threads. Another important advantage is that the code can be compiled as a single-threaded application just disabling support for OpenMP, thus easing debugging and so programmability.

Overall, the use of OpenMP presents three main advantages. First, an expert community has constantly reviewed and augmented the language for the past 20 years. Second, OpenMP is widely implemented by several chip and compiler vendors from both high-performance and embedded computing domains (e.g., GNU, Intel®, ARM, Texas Instruments and IBM), increasing portability among multiple platforms from different vendors. Third, OpenMP provides greater expressiveness due to years of experience in its development; the language offers several directives for parallelization and fine-grained synchronization, along with a large number of clauses that allow it to contextualize concurrency and heterogeneity, providing fine control of the parallelism.

3.2 The OpenMP Parallel Programming Model

3.2.1 Introduction and Evolution of OpenMP

OpenMP represents the computing resources of a parallel processor architecture (i.e., cores) by means of high-level threads, named *OpenMP threads*, upon which programmers can assign units of code to be executed. During the execution of the program, the OpenMP runtime assigns these threads to low-level computing resources, i.e., the operating system (OS) threads, which are then assigned to physical cores by the OS scheduler, following the execution model defined by the OpenMP directives. Figure 3.1 shows a schematic view of the stack of components involved in the execution of an OpenMP program. OpenMP exposes some aspects of managing OpenMP threads to the user (e.g., defining the number of OpenMP threads assigned to a parallel execution by means of the num_threads clause). The rest of components are transparent to the user and efficiently managed by the OpenMP runtime and the OS.

Originally, up to OpenMP version 2.5 [22], OpenMP was traditionally focused on massively data-parallel, loop-intensive applications, following the *single-program-multiple-data* programming paradigm. In this model, known as *thread model*, OpenMP threads are visible to the programmer, which are controlled with work-sharing constructs that assign iterations of a loop or code segments to OpenMP threads.

The OpenMP 3.0 specification [23] introduced the concept of tasks by means of the task directive, which exposes a higher level of abstraction to programmers. A task is an independent parallel unit of work, which defines an instance of code and its data environment. This new model, known as *tasking model*, provides a very convenient abstraction of parallelism as it is the runtime (and not the programmer) the responsible for scheduling tasks to threads.

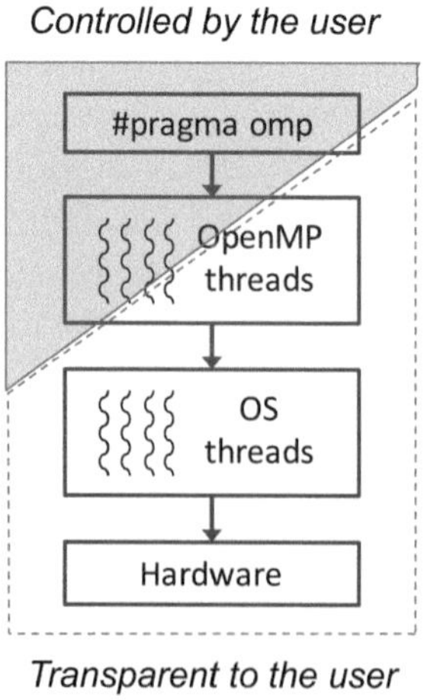

Figure 3.1 OpenMP components stack.

With version 4.0 of the specification [24], OpenMP evolved to consider very sophisticated types of fine-grained, irregular and highly unstructured parallelism, with mature support to express dependences among tasks. Moreover, it incorporated for the first time a new accelerator model including features for offloading computation and performing data transfers between the host and one or more accelerator devices. The latest version, OpenMP 4.5 [25], enhances the previous accelerator model by coupling it with the tasking model.

Figure 3.2 shows a time-line of all existent releases of OpenMP, since 1997, when the OpenMP Architecture Review Board (ARB) was formed. The next version, 5.0 [26–28], is planned for November 2018.

3.2.2 Parallel Model of OpenMP

This section provides a brief description of the OpenMP parallel programming model as defined in the latest specification, version 4.5.

3.2.2.1 Execution model

An OpenMP program begins as a single thread of execution, called the *initial thread*. Parallelism is achieved through the parallel construct, in which a new *team* of OpenMP threads is spawned. OpenMP allows programmers to define the amount of threads desired for a parallel region by means of the num_threads clause attached to the parallel construct. The spawned threads are joined at the implicit barrier encountered at the end of the parallel region. This is the so-called *fork-join model*. Within the parallel region, parallelism can be distributed in two ways that provide tantamount performance [29]:

1. The *thread-centric model* exploits structured parallelism distributing work by means of work-sharing constructs (e.g., for and sections constructs). It provides a fine-grained control of the mapping between

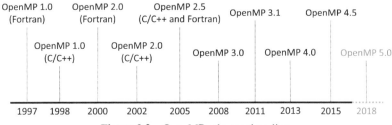

Figure 3.2 OpenMP releases time-line.

work and threads, as well as a coarse grain synchronization mechanism by means of the barrier construct.

2. The *task-centric model*, or simply *tasking model*, exploits both structured and unstructured parallelism distributing work by means of tasking constructs (e.g., task and taskloop constructs). It provides a higher level of abstraction in which threads are mainly controlled by the runtime, as well as fine-grained synchronization mechanisms by means of the taskwait construct and the depend clause that, attached to a task construct, allow the description of a list of *input* and/or *output* dependences. A task with an in, out or inout dependence is ready to execute when all previous tasks with an out or inout dependence on the same storage location complete.

Figure 3.3 shows the execution model of a parallel loop implemented with the for directive, where all spawned threads work in parallel from the beginning of the parallel region as long as there is work to do. Figure 3.4 shows the model of a parallel block with unstructured tasks. In this case, the single construct restricts the execution of the parallel region to only one thread until a task construct is found. Then, another thread (or the same, depending on the scheduling policy), concurrently executes the code of the task. In Figure 3.3, the colours represent the execution of differents iterations of the same parallel loop; in Figure 3.4, colours represent the parallel execution of the code included within a task construct.

3.2.2.2 Acceleration model
OpenMP also provides a *host-centric accelerator model* in which a host offloads data and code to the accelerator devices available in the same

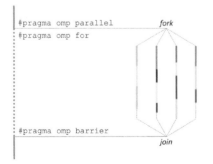

Figure 3.3   Structured parallelism.

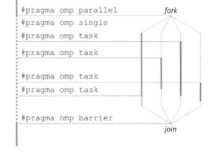

Figure 3.4   Unstructured parallelism.

processor architecture for execution by means of the target construct. When a target directive is encountered, a new *target task* enclosing the target region is generated. The target task is completed after the execution of the target region finishes. One of the most interesting characteristics of the accelerator model is its integration with the tasking model. Note that each accelerator device has its own team of threads that are distinct from threads that execute on another device, and these cannot migrate from one device to another.

In case the accelerator device is not available or even does not exist (this may occur when the code is ported from one architecture to another) the target region is executed in the host. The map clause associated with the target construct specifies the data items that will be mapped to/from the target device. Further parallelism can be exploited within the target device.

3.2.2.3 Memory model

OpenMP is based on a relaxed-consistency, shared-memory model. This means there is a memory space shared for all threads, called *memory*. Additionally, each thread has a temporary view of the memory. The temporary view is not always required to be consistent with the memory. Instead, each private view synchronizes with the main memory by means of the *flush* operation, which can be implicit (due to operations causing a memory fence) or explicit (using the flush operation). Data cannot be directly synchronized between two different threads temporary view.

The view of each thread has of a given variable is defined using data-sharing clauses, which can determine the following sharing scopes:

- private: a new fresh variable is created within the scope.
- firstprivate: a new variable is created in the scope and initialized with the value of the original variable.
- lastprivate: a new variable is created within the scope and the original variable is updated at the end of the execution of the region (only for tasks).
- shared: the original variable is used in the scope, thus opening the possibility of data race conditions.

The use of data-sharing clauses is particularly powerful to avoid unnecessary synchronizations as well as race conditions. All variables appearing within a construct have a default data-sharing defined by the OpenMP specification ([25] Section 2.15.1). These rules are not based on the use of the variables, but on their storage. Thus, users are duty-bound to explicitly scope many variables, changing the default data-sharing values, in order to fulfill correctness (e.g., avoiding data races) and enhance performance (e.g., avoiding unnecessary privatizations).

3.2.3 An OpenMP Example

Listing 3.1 illustrates an OpenMP program that uses both the tasking and the accelerator models. The code enclosed in the parallel construct (line 4) defines a team of four OpenMP threads on the host device. The single construct (line 6) specifies that only one thread starts executing the associated block of code, while the rest of threads in the team remain waiting. When the task regions are created (lines 9 and 11), each one is assigned to one thread in the team (may be the same thread), and the corresponding output dependences on variables x and y are stored. When the target task (lines 13:14) is created, its dependences on x and y are checked. If the tasks producing these variables are finished, then the target task can be scheduled. Otherwise, it must be deferred until the tasks from which it depends have finished. When the target task is scheduled, the code contained in the target region and the variables in the map(to:) clause (x and y) are copied to the accelerator device. After its execution, the res variable is copied back to the host memory as defined by the map(from:) clause. The presence of a nowait clause in the target task allows the execution on the host to continue after the target task is created.

Listing 3.1   OpenMP example of the tasking and the accelerator models combined

```
1  int foo(int a, int b)
2  {
3      int res;
4      #pragma omp parallel num_threads(4) shared(res) firstprivate(a, b)
5      {
6      #pragma omp single shared(res) firstprivate(a, b)
7      {
8          int x, y;
9          #pragma omp task shared(x) firstprivate(a) depend(out:x)
10         x = a*a;
11         #pragma omp task shared(y) firstprivate(b) depend(out:y)
12         y = b*b;
13         #pragma omp target map(to:x,y) map(from:res) nowait \
14                            shared(res) firstprivate(x, y) depend(in:x,y)
15         res = x + y;
16     }
17     return res;
18     }
19 }
```

All OpenMP threads are guaranteed to be synchronized at the implicit barrier included at the end of the parallel and single constructs (lines 16 and 19 respectively). A nowait clause could be added to the single construct to avoid unnecessary synchronizations.

3.3 Timing Properties of the OpenMP Tasking Model

The tasking model of OpenMP not only provides a very convenient abstraction layer upon which programmers can efficiently develop parallel applications, but also has certain similarities with the *sporadic direct acyclic graph (DAG) scheduling model* used to derive a (worst-case) response time analysis of parallel applications. Chapter 4 presents in detail the response time analyses that can be applied to the OpenMP tasking model. This section derives the OpenMP-DAG upon which these analyses are applied.

3.3.1 Sporadic DAG Scheduling Model of Parallel Applications

Real-time embedded systems are often composed of a collection of periodic processing stages applied on different input data streaming coming from sensors. Such a structure makes the system amenable to timing analysis methods [30].

The *task model* [31], either *sporadic or periodic*, is a well-known model in scheduling theory to represent real-time systems. In this model, real-time applications are typically represented as a set of n *recurrent* tasks $\tau = \{\tau_1, \tau_2, .., \tau_n\}$, each characterized by three parameters: worst-case execution time ($WCET$), period (T) and relative deadline (D). Tasks repeatedly emit an infinite sequence of *jobs*. In case of periodic tasks, jobs arrive strictly periodically separated by the fixed interval time T. In case of sporadic tasks, jobs do not have a strict arrival time, but it is assumed that a new job released at time t must finish before $t + D$. Moreover, a minimum interval of time T must occur between two consecutive jobs from the same task.

With the introduction of multi-core processors, new scheduling models have been proposed to better express the parallelism that these architectures offer. This is the case of the *sporadic DAG task model* [32–36], which allows the exploitation of parallelism *within* tasks. In the sporadic DAG task model each task (called *DAG-task*) is represented with a *directed acyclic graph* (DAG) $G = (V, E)$, T and D. Each node $v \in V$ denotes a sequential operation characterized by a WCET estimation. Edges represent dependences between nodes: if $e = (v_1, v_2) : e \in E$, then the node v_1 must complete its execution before node v_2 can start executing. In other words, the DAG captures scheduling constraints imposed by dependences among nodes and it is annotated with the WCET estimation of each individual node.

Overall, the DAG represents the main formalism to capture the properties of a real-time application. In that context, although the current specification of OpenMP lacks any notion of real-time scheduling semantics, such as deadline, period, or WCET, the structure and syntax of an OpenMP program have certain similarities with the DAG model. The task and taskwait constructs, together with the depend clause, are very convenient for describing a DAG. Intuitively, a task describes a node in V in the DAG model, while taskwait constructs and depend clauses describe the *edges* in E. Unfortunately, such a DAG would not convey proper information to derive a real-time schedule that complies with the semantics of the OpenMP specification.

In order to understand where the difficulties of mapping an OpenMP program onto an expressive task graph stem from, and how to overcome them, the next section further delves into the details of the OpenMP execution model.

3.3.2 Understanding the OpenMP Tasking Model

When a task construct is encountered, the execution of the new task region can be assigned to one of the threads in the current team for immediate or deferred execution, with the corresponding impact on the overall timing behaviour. Different clauses allow defining how a task, its parent task and its child tasks will behave at runtime:

- The depend clause allows describing a list of input (in), output (out), or input-output (inout) dependences on data items. Dependences can only be defined among *sibling* tasks, i.e., first-level descendants of the same *parent* task.
- An if clause whose associated expression evaluates to false forces the encountering thread to suspend the current task region. Its execution cannot be resumed until the newly generated task, defined to be an *undeferred task*, is completed.
- A final clause whose associated expression evaluates to true forces all its child tasks to be *undeferred* and *included* tasks, meaning that the encountering thread itself sequentially executes all the new descendants.
- By default, OpenMP tasks are *tied* to the thread that first starts their execution. If such tasks are suspended, they can only be resumed by the same thread. An untied clause forces the task not to be tied to any thread; hence, in case it is suspended, it can later be resumed by any thread in the current team.

Listing 3.2 OpenMP example of task scheduling clauses

```
1  #pragma omp parallel
2  {
3  #pragma omp single nowait                              // T₀
4  {
5          ...              // tp₀₀
6      #pragma omp task depend(out:x) untied final(true)   // T₁
7      {
8              ...          // tp₁₀
9          #pragma omp task                                // T₄
10         { ... } // tp₄
11             ...          // tp₁₁
12     }
13         ...              // tp₀₁
14     #pragma omp task depend(in:x)                       // T₂
15     { ... }     // tp₂
16         ...              // tp₀₂
17     #pragma omp taskwait
18         ...              // tp₀₃
19     #pragma omp task                                    // T₃
20     { ... }     // tp₃
21         ...              // tp₀₄
22 }
23 }
```

Listing 3.2 shows an example of an OpenMP program using different tasking features. The parallel construct creates a new team of threads (since num_threads clause is not provided, the number of threads associated is implementation defined). The single construct (line 3) generates a new task region T_0, and its execution is assigned to just one thread in the team. When the thread executing T_0 encounters its child task constructs (lines 6, 14, and 19), new tasks T_1, T_2, and T_3 are generated. Similarly, the thread executing T_1 creates task T_4 (line 9).

Tasks T_1 and T_2 include a depend clause both defining a dependence on the memory reference x, so T_2 cannot start executing until T_1 finishes. T_4 is defined as an *included task* because its parent T_1 contains a final clause that evaluates to *true*, so T_1 is suspended until the execution of T_4 finishes. All tasks are guaranteed to have completed at the *implicit barrier* at the end of the parallel region (line 23). Moreover, task T_0 will wait on the taskwait (line 17) until tasks T_1 and T_2 have completed before proceeding.

OpenMP defines *task scheduling points* (TSPs) as points in the program where the encountering task can be suspended, and the hosting thread can be rescheduled to a different task. TSPs occur upon task creation and completion and at task synchronization points such as taskwait directives or explicit and implicit barriers[1].

Task scheduling points divide task regions into *task parts* executed uninterruptedly from start to end. Different parts of the same task region are executed in the order in which they are encountered. In the absence of task synchronization constructs, the order in which a thread executes parts of different tasks is unspecified. The example shown in Figure 3.2 identifies the parts in which each task region is divided: T_0 is composed of task parts $tp_{00}, tp_{01}, tp_{02}, tp_{03}$, and tp_{04}; T_1 is composed of task parts tp_{10}, and tp_{11}; and T_2, T_3, and T_4 are composed of task part tp_2, tp_3, and tp_4, respectively.

When a task encounters a TSP, the OpenMP runtime system may either begin the execution of a task region bound to the current team, or resume any previously suspended task region also bound to it. The order in which these actions are applied is not specified by the standard, but it is subject to the following *task scheduling constraints* (TSCs):

TSC 1: An *included* task must be executed immediately after the task is created.

TSC 2: Scheduling of new *tied* tasks is constrained by the set of task regions that are currently tied to the thread, and that are not suspended in a barrier region. If this set is empty, any new *tied* task may be scheduled. Otherwise, a new *tied* task may be scheduled only if all tasks in the set belong to the same task region and the new *tied* task is a *child task* of the task region.

TSC 3: A dependent task shall not be scheduled until its task data dependences are fulfilled.

TSC 4: When a task is generated by a construct containing an if clause for which the conditional expression evaluates to false, and the previous constraints are already met, the task is executed immediately after generation of the task.

3.3.3 OpenMP and Timing Predictability

The execution model of OpenMP tasks differs from the DAG model in a fundamental aspect: a node in the DAG model is a sequential operation that

---

[1]Additional TSPs are implied at different OpenMP constructs (target, taskyield, taskgroup). See Section 2.9.5 of the OpenMP specification [25] for a complete list of task scheduling points.

cannot be interrupted[2]. Instead, an OpenMP task can legally contain multiple TSPs at which the task can be suspended or resumed following the TSCs.

Moreover, in order to correctly capture scheduling constraints of each task as defined by the OpenMP specification, a DAG-based real-time scheduling model requires to know: (1) the dependences among tasks, (2) the point in time of each TSP, and (3) the scheduling clauses associated to the task.

This section analyses the extraction of a DAG that represents the parallel execution of an OpenMP application upon which timing analysis can be then applied. It focuses on three key elements:

1. How to reconstruct an OpenMP task graph from the analysis of the code that resembles the DAG-task structure based on TSPs.
2. To which elements of an OpenMP program WCET analysis must be applied.
3. How to schedule OpenMP tasks based on DAG-task methodologies so that TSCs are met.

3.3.3.1 Extracting the DAG of an OpenMP program

The execution of a task part resembles the execution of a node in V, i.e., it is executed uninterrupted. To that end, OpenMP task parts, instead of tasks, can be considered as nodes in V.

Figure 3.5 shows the DAG (named *OpenMP-DAG*) corresponding to the example presented in Listing 3.2, in which task parts form the nodes in V. T_0 is decomposed into task parts tp_{00}, tp_{01}, tp_{02}, tp_{03}, and tp_{04}, with a TSP at the end of each part caused by the task constructs T_1, T_2, and T_3 for tp_{00}, tp_{01}, and tp_{03}, and the taskwait construct for tp_{02}. Similarly, T_1 is decomposed into tp_{10} and tp_{11} with the TSP corresponding to the creation of task T_4 at the end of tp_{10}.

Depending on the origin of the TSP encountered at the end of each task part (i.e., task creation or completion, or task synchronization) three different types of dependences are identified: (a) control-flow dependences (dotted arrows), which force parts to be scheduled in the same order as they are executed within the task; (b) TSP dependences (dashed arrows), which force tasks to start/resume execution after the corresponding TSP, and (c) full synchronizations (solid arrows), which force the sequential execution of tasks as defined by the depend clause and task synchronization constructs. Note that all dependence types have the same purpose, which is to express

---

[2]This assumes the execution of a single DAG program, where a node cannot be interrupted to execute other nodes of the same graph. In a multi-DAG execution model, nodes can be preempted by nodes from different DAG programs if allowed by the scheduling approach.

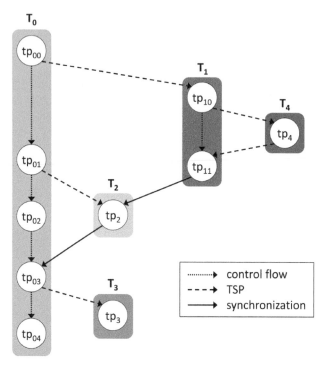

Figure 3.5   OpenMP-DAG composed of *task parts* based on the code in Listing 3.2.

a scheduling precedence constraint. As a result, the OpenMP-DAG does not require to differentiate them.

Besides the depend clause, the if and final clauses also affect the order in which task parts are executed. In both cases the encountering task is suspended until the newly generated task completes execution. In order to model the undeferred and included tasks behaviour, a new edge is introduced in E. In Figure 3.5, a new dependence between tp_{40} and tp_{11} is inserted, so the task region T_1 does not resume its execution until the included task T_4 finishes.

3.3.3.2 WCET analysis is applied to *tasks* and *task parts*
In order to comply with the DAG-model, nodes in the OpenMP-DAG must be further annotated with the WCET estimation of the corresponding task parts. By constructing the OpenMP-DAG based on the knowledge of TSPs (i.e., by considering as nodes in V only those code portions that are executed uninterruptedly from start to end) the timing analysis of each node has

a WCET which is independent of any dynamic instance of the OpenMP program (i.e., how threads may be scheduled to tasks and parts therein). As a result, the timing behaviour of task parts will only be affected by concurrent accesses to shared resources [37]. It is important to remark that the WCET estimation is applied to a task when it is composed of a single task part. This is the case of T_2, T_3, and T_4 from Figure 3.5.

3.3.3.3 DAG-based scheduling must not violate the TSCs

When real-time scheduling techniques are applied to guarantee the timing behaviour of OpenMP applications, the semantics specified by the OpenMP TSCs must not be violated.

The clauses associated to a task construct not only define precedence constraints, as shown in Section 3.3.3.1, but they also define the way in which tasks, and task parts therein, are scheduled according to the TSCs defined in Section 3.3.2. This is the case of the if, final and untied clauses, as well as the default behaviour of *tied* tasks. These clauses influence the order in which tasks execute and also how task parts are scheduled to threads. Regarding the latter, the restrictions imposed by TSCs are the following:

- *TSC 1* imposes *included* tasks to be executed immediately by the encountering thread. In this case, the scheduling of the OpenMP-DAG has to consider both the task part that encounters it and the complete *included* task region as a unique unit of scheduling. In Figure 3.5, the former case would give tp_4 the highest priority, and the latter case would consider tp_{10} and tp_4 as a unique unit of scheduling.
- *TSC 2* does not allow scheduling new *tied* tasks if there are other suspended *tied* tasks already assigned to the same thread, and the suspended tasks are not descendants of the new task. Listing 3.3 shows a fragment of code in which this situation can occur. Let's assume that T_1, which is not a descendent of T_3, is executed by *thread 1*. When T_1 encounters

---

Listing 3.3   Example of an OpenMP fragment of code with *tied tasks*

```
1  ...
2  #pragma omp task  // T₁
3  {
4       #pragma omp task if(false)  // T₂
5       {...}
6  }
7  #pragma omp task  // T₃
8  {...}
```

the TSP of the creation of T_2, it is suspended because of TSC 4, and it cannot resume until T_2 finishes. Let's consider that T_2 is being executed by a different thread, e.g., *thread 2*. If T_2 has not finished when the TSP of the creation of T_3 is reached, then T_3 cannot be scheduled on *thread 1* because of TSC 2, even if *thread 1* is idle. As a result, *tied* tasks constrain the scheduling opportunities of the OpenMP-DAG.

- *TSC 3* imposes tasks to be scheduled respecting their dependences. This information is already contained in the OpenMP-DAG.
- *TSC 4* states that *undeferred* tasks execute immediately if TSCs 1, 2, and 3 are met. Differently, *untied* tasks are not subject to any TSC, allowing parts of the same task to execute on different threads, so when a task is suspended, the next part to be executed can be resumed on a different thread. Therefore, one possible scheduling strategy for *untied* tasks that satisfies TSC 4 is not to schedule *undeferred* and *untied* task parts until *tied* and *included* tasks are assigned to a given thread. This guarantees that TSCs 1 and 2 are met. This is because task parts of *tied* and *included* tasks are bound to the thread that first started their execution, which reduces significantly their scheduling opportunities. Instead, *untied* and *undeferred* task parts have a higher degree of freedom as they can be scheduled to any thread of the team. Therefore, for the OpenMP-DAG to convey enough information to devise a TSC-compliant scheduling, each node in V must be augmented with the *type of task* as well (untied, *tied*, *undeferred* and *included*) as shown in Figure 3.5.

Figure 3.6 shows a possible schedule of task parts in Listing 3.2, assuming a work-conserving scheduling. T_0 is a tied task, so all its task parts are scheduled to the same thread (*thread 1*). T_1 is an *untied* task so tp_{10} and tp_{10} can execute in different threads (*thread 1* and *2* in the example). Note that tp_{11} does not start executing until tp_4 completes due to the TSP constraint. Moreover, the execution of tp_4 starts immediately after the creation of T_4 on

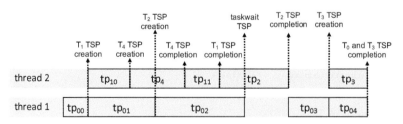

Figure 3.6   DAG composed of task parts.

the same thread that encounters it (*thread 2*). Finally, tp_2 and tp_3 are scheduled to idle threads (*thread 4* and *5*, respectively) once all their dependences are fulfilled.

3.4 Extracting the Timing Information of an OpenMP Program

The extraction of an OpenMP-DAG representing the parallel execution of an OpenMP program in such a way that timing analysis can be performed, requires analyzing the OpenMP constructs included in the source code, so the nodes and edges that form the DAG can be identified. This information can be obtained by means of compiler analysis techniques. Concretely, there exists two different analysis stages needed to build the OpenMP-DAG $G = (V, E)$:

1. A *parallel structure stage*, in which the nodes in V, i.e., tasks parts, and edges in E, are identified based on TSPs, TSCs, and data- and control-flow information.

Listing 3.4 OpenMP program using the tasking model

```
1  #pragma omp parallel num_threads(8)
2  {
3  #pragma omp single nowait                              //T₀
4  {
5      for(i=0; i<=2; i++)
6        for(int j=0; j<=2; j++) {
7          if(i==0 && j==0) {              // Initial block
8            #pragma omp task depend(inout:m[i][j])
9            compute_block(i, j);                    // T₁
10         } else if (i == 0) {      // Blocks in upper edge
11           #pragma omp task depend(in:m[i][j−1], inout:m[i][j])
12           compute_block(i, j);               // T₂
13         } else if (j == 0) {       // Blocks in left edge
14           #pragma omp task depend(in:m[i−1][j], inout:m[i][j])
15           compute_block(i, j);               // T₃
16         } else {                    // Internal blocks
17           #pragma omp task depend(in:m[i−1][j], in:m[i][j−1], \\
18                                   in:m[i−1][j−1], inout:m[i][j])
19           compute_block(i, j);               // T₄
20         }
21       }
22  }
23  }
24  }
```

2. A *task expansion stage*, in which the tasks (and task parts) that will be actually instantiated at runtime are identified by expanding the control flow information extracted in the previous stage.

The next subsections further describe these stages. With the objective of facilitating the explanation of the compiler analysis techniques, Listing 3.4 introduces an OpenMP program that will be used for illustration purposes. The code processes the elements of a blocked 2D matrix using a *wave-front* parallelization strategy [38]. The parallel construct (line 1) defines a team of 8 threads. The single construct (line 3) specifies that only one thread will execute the associated code. The algorithm divides the matrix in 3×3 blocks, assigning each one to a different task. Each block $[i, j]$ consumes the previous adjacent blocks and itself. Hence, all tasks (lines 8, 11, 14, and 17:18) have an inout dependence on the computed block $[i, j]$. T_2 and T_3 (lines 11 and 14) compute the upper and left edges, so additionally they consume the left $[i, j - 1]$ and upper $[i - 1, j]$ blocks, respectively. Finally, T_4 (lines 17:18) computes the internal blocks, hence additionally it consumes the left $[i - 1, j]$, upper $[i, j - 1]$, and left-upper diagonal $[i - 1, j - 1]$ blocks. All tasks are guaranteed to complete at the implicit barrier at the end of the parallel region (line 24).

3.4.1 Parallel Structure Stage

This stage identifies the TSPs surrounding tasks parts, and the corresponding TSCs associated with each task part in order to: (1) generate a parallel control-flow graph (PCFG) that holds all this information as well as parallel semantics [39], and (2) analyze this graph so that the necessary information to expand a complete DAG is obtained. With such purpose in mind the analysis performs the following calculations:

- Generate the PCFG of the source code taking into account: (a) the dependences introduced by any kind of TSPs (i.e., task creation, task completion and task synchronization), as introduced in Section 3.3.3.1, (b) the data dependences introduced by the depend clause, and (c) the if and final clauses, hence the behaviour of undeferred and included tasks.
- On top of that, analyze the control-flow statements, i.e., selection statements (if-else and switch) and loops that identify whether a task is instantiated or not at runtime. To do so, three analyses are required: induction variables [40], reaching definitions [41], and range analysis [42]. Additionally, determine the conditions that must be fulfilled for two instantiated tasks to depend on one another [3].

3.4.1.1 Parallel control flow analysis

The *abstract syntax tree* (AST) used in the compiler to represent the source code is used to generate the PCFG of an OpenMP program. This enriches the classic control-flow graph (CFG) with information about parallel execution. This process performs a conservative analysis of the synchronizations among tasks, because the compiler may not be able to assert when two depend clauses designate the same memory location, e.g., array accesses or pointers. Hence, synchronization edges are augmented with predicates defining the condition to be fulfilled for an edge to exist. In the example shown in Listing 3.4, the dependences that matrix m originates among tasks depend on the values of i and j.

3.4.1.2 Induction variables analysis

On top of the PCFG, the compiler evaluates the loop statements to discover the induction variables (IVs) and their evolution over the iterations using the common tuple representation $\langle lb, ub, str \rangle$, where lb is the lower bound, ub is the upper bound, and str is the stride. This analysis is essential for the later expansion of the graph, since the induction variables will determine the shape of the iteration space for each loop statement.

3.4.1.3 Reaching definitions and range analysis

Finally, the compiler computes the values of all variables involved in the execution of any task. With such a purpose, it analyzes reaching definitions and also extends range analysis with support for OpenMP. The former computes the definitions reaching any point in the program. The later computes the values of the variables at any point of the program in four steps: (1) generate a set $\mathcal{C}$ of equations that constrain the values of each variable (equations are built for each assignment and control flow statement); (2) build a *constraint graph* that represents the relations among the constraints; (3) split the graph into *strongly connected components* (SCCs) to avoid cycles; (4) propagate the ranges over the SCCs in topological order. Both analyses are needed to propagate the values of the relevant variables across the expanded code.

3.4.1.4 Putting all together: The wave-front example

The previously mentioned analyses provide the information needed to generate an initial version of the DAG, named *augmented DAG (aDAG)*, with data and control flow knowledge. The aDAG is defined by the tuple

$$aDAG = \langle N, E, C \rangle \tag{3.1}$$

where:

- $N = \{V \times T_N\}$ is the set of nodes with their corresponding type $T_N = \{Task, Taskwait, Barrier\}$.
- $E = \{N \times N \times P\}$ is the set of possible synchronization edges with the predicate P that must fulfill for the edge to exist.
- $C = N \times \{F\}$ is the set of *control flow statements* involved in the instantiation of any task $n \in N$, where $F = S \times \{T_F\}$, being S the condition to instantiate the tasks and $T_F = \{Loop, IfElse, Switch\}$, the type of the structure.

Figure 3.7 shows the aDAG of the OpenMP program in Listing 3.4. The set of nodes N includes all task constructs $N = T_1, T_2, T_3, T_4$ (lines 8, 11, 14, and 17:18), all with type $T_N = Task$. The control flow statements for each node N, $f_i \in F$ are the for (lines 5 and 6) and if (lines 7, 10, 13, and 16) statements, and include information about: (a) the IVs of each loop i, j, both with $lb = 0$, $ub = 2$ and $str = 1$ (dashed-line boxes); (b) the conditions of the selection statements enclosing each task (solid-line boxes), and (c) the ranges of the variables in those conditions. In the figure, T_3 is instantiated if $i = 1$ or 2 and $j = 0$. In the predicates $p \in P$ associated to the synchronization edges in E, the left hand side of the equality corresponds to the value of the variable at the point in time the source task is instantiated, while the right side corresponds to the value when the target task is instantiated. For example, the predicate of the edge between T_1 and T_3 with $p1((i_S == i_T || i_S == i_T - 1)\&\&j_S == j_T)$ evaluates to *true*, meaning that the edge exists when the values of i and j in the source task T_1 are $i_S = 0$ and $j_S = 0$, and the values of i and j in the target task T_3 are $i_T = 1$ and $j_T = 0$.

For simplicity, Figure 3.7 only includes the dependences that are actually expanded in the next stage (Section 3.4.2). The actual aDAG has edges between any possible pair of tasks because they all have *inout* dependences on the element $m[i][j]$. Moreover, the task-parts that form the task T_0 with the corresponding task creation dependences are not included.

3.4.2 Task Expansion Stage

3.4.2.1 Control flow expansion and synchronization predicate resolution

Based on the aDAG, this stage generates an *expanded DAG* (or simply DAG) representing the complete execution of the program in two phases: (1) expand control flow structures (i.e., decide which branches are taken for the selection

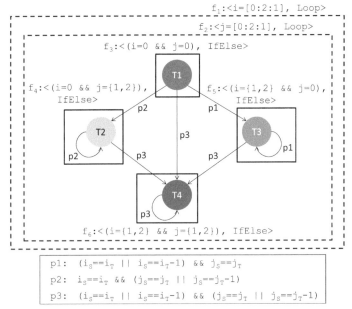

```
p1:  (i_S==i_T || i_S==i_T-1) && j_S==j_T
p2:  i_S==i_T && (j_S==j_T || j_S==j_T-1)
p3:  (i_S==i_T || i_S==i_T-1) && (j_S==j_T || j_S==j_T-1)
```

Figure 3.7 aDAG of the OpenMP program in Listing 3.4.

statements and how many iterations are executed for the loop statements) to determine which tasks (and so task-parts) are actually instantiated; and (2) resolve the synchronization predicates to conclude which tasks have actual dependences.

Control flow structures are expanded from outer to inner levels. In the aDAG in Figure 3.7, the outer loop f_1 is expanded first, and then the inner loop f_2. Finally, the if-else structures f_3, f_4, f_5, and f_6 are resolved. Each expansion requires the evaluation of the associated expressions to determine the values of each variable. For example, when the outer loop f_1 is expanded, each iteration is associated with the corresponding value of i.

This expansion process creates two identifiers: (1) an identifier of the loops involved in the creation of a task (l_i), labeling each loop expansion step, and (2) a unique *static task construct identifier* (sid_t), labeling each task construct.

The process results in a temporary DAG in which all tasks instantiated at runtime are defined, but synchronization predicates are not solved. To do so, the value of the variables propagated in the control flow expansion is used to evaluate predicates and decide which edges actually exist.

Likewise, loop identifiers l_i are used to eliminate backwards dependences, i.e., tasks instantiated in previous iterations cannot depend on tasks instantiated in later iterations.

Figure 3.8 shows the final DAG of the program in Listing 3.4. It contains all task instances with a unique numerial identifier (explained in the next section) and all dependences that can potentially exist at runtime. Transitive dependences (dashed arrows) are included as well, although they can be removed because they are redundant.

3.4.2.2 t_{id}: A unique task instance identifier

A key property of the expanded task instances is that they must include a *unique task instance identifier* t_{id} required to match the instantiated tasks expanded at compile-time (and included in the DAG) with those instantiated at runtime. Equation 3.2 computes t_{id} as follows:

$$t_{id} = sid_t + T \times \sum_{i=1}^{L_t} l_i \cdot M^i \tag{3.2}$$

where sid_t is a unique task construct identifier (computed during the control flow expansion stage), T is equal to the number of task, taskwait, and barrier constructs in the source code, L_t is the total number of nested loops involved in the execution of the task t, i refers to the the nesting level, l_i is the loop unique identifier at nesting level i (computed during the control

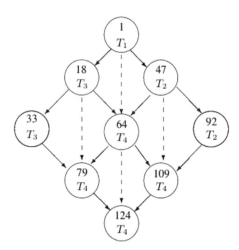

Figure 3.8   The DAG of the OpenMP program in Listing 3.4.

flow expansion stage), and M the maximum number of iterations of any considered loop.

The use of loop properties in Equation 3.2 (i.e., L_t, l_i, i, and M), guarantees that a unique task identifier for each task instance is generated, even if they come from the same task construct. Hence, task instances from different loop iterations result in different t_{id} because every nesting level l_i is multiplied by the maximum number of iterations M.

Consider task T_4, with identifier 79, in Figure 3.8. This task instance corresponds to the computation of the matrix block $m[2, 1]$. Its identifier is computed as follows: (1) $sid_{T4} = 4$, because T_4 is the fourth task found in sequential order while traversing the source code; (2) $T = 5$ because there are four task constructs and one (implicit) barrier in the source code; (3) $L_{T_4} = 2$, the two nested loops enclosing T_4; (4) $M = 3$, the maximum number of iterations in any of the two considered loops; and (5) $l_1 = 2$ and $l_2 = 1$ are the values of the loop identifiers at the corresponding iteration. Putting them all together: $T_{4_{id}} = 4 + 5(2 * 3^1 + 1 * 3^2) = 79$.

It is important to remark that t_{id} must be computed at both compile-time and run-time, and so all information needed to compute Equation 3.2 must be available in both places. Chapter 6 presents the combined compiler and run-time mechanisms needed to reproduce all the required information (including sid_t and l_i identifiers) at run-time.

3.4.2.3 Missing information when deriving the DAG

In case the framework cannot derive some information (mostly when control-flow statements and dependences contain pointers that may alias or arrays with unresolved subscripts, or the values are not known at compile-time), it still generates a DAG that correctly represents the execution of the program. Next, each possible case is argued:

- When an if-else statement cannot be evaluated, all its related tasks in C are considered for instantiation, hence included in the DAG. In this case, the DAG will include a task instance that will never exist. Chapters 4 and 6 present the mechanisms required to take this into consideration for response time analysis and parallel run-time execution.
- If a loop cannot be expanded because its boundaries are unknown, parallelism across iterations is disabled by inserting a taskwait at the end of the loop. By doing so, all tasks instantiated within an iteration must complete before the next iteration starts.

- Lastly, dependences whose predicate cannot be evaluated are always kept in the DAG, making the involved tasks serialized.

The situations described above will result in a bigger DAG (when if–else conditions cannot be evaluated) or in a performance loss (when loop bounds or synchronization predicates cannot be determined), although a correct DAG is guaranteed. In the worst-case scenario, where no information can be derived at compile-time, the resultant DAG corresponds to the sequential execution of the program, i.e., all tasks are assumed to be instantiated, and their execution is to be sequentialized. It is important to remark that embedded applications can often provide all the required information to complete the DAG expansion, as it is required for timing analysis [43].

3.4.3 Compiler Complexity

The complexity of the compiler is determined by the complexity of the two stages presented in Sections 3.4.1 and 3.4.2.

The complexity of the control/data flow analysis stage is dominated by the PCFG analysis and range analysis phases. The complexity of the former is related to the number of split constructs present in the source code, in which the Cyclomatic Complexity [44] metric is usually used. The latter, has been proved to have an asymptotic linear complexity [42].

The complexity of the task expansion stage is dominated by the computation of the dependences among tasks, which is performed using a Cartesian product: the input dependence of a task can be generated by any of the previously created task instances. As a result, the complexity is quadratic on the number of instantiated tasks.

3.5 Summary

This chapter provided the rationale and the model for the use of fine-grained parallelism in general, and the OpenMP parallel programming model in particular, to support applications that require predictable performance, to develop future critical real-time embedded systems, and analyze the time predictable properties of the OpenMP tasking model. Based on this model, the chapter then described the advances in compiler techniques to extract timing information of OpenMP parallel programs, and build the *OpenMP* DAG required to enable predictable scheduling (described in the next chapter) and the needed timing analysis (in Chapter 5). This OpenMP-DAG also provides the building block for the execution of the OpenMP runtime (Chapter 6) and Operating System (Chapter 7).

References

[1] Pllana, S., and Xhafa, F., *Programming Multicore and Many-core Computing Systems*, volume 86. John Wiley and Sons, 2017.

[2] Reinders, J., *Intel Threading Building Blocks*. O'Reilly and Associates, Inc., 2007.

[3] *NVIDIA CUDA C Programming Guide*. https://docs.nvidia.com/cuda /cuda-c-programming-guide/index.html, 2016.

[4] Stone, J. E., Gohara, D., and Shi, G., OpenCL: A parallel programming standard for heterogeneous computing systems. *CSE*, 12, 66–73, 2010.

[5] Snir, M., *MPI–the Complete Reference: The MPI core*, volume 1. MIT press, 1998.

[6] Butenhof, D. R., *Programming with POSIX Threads*. Addison-Wesley, 1997.

[7] Chapman, B., Jost, G., and Van Der Pas., *Using OpenMP: Portable Shared Memory Parallel Programming*, volume 10. MIT press, 2008.

[8] Duran, A., Ayguadé, E., Badia, R. M., Labarta, J., Martinell, L., Martorell, X., and Planas, J., Ompss: a proposal for programming heterogeneous multi-core architectures. *Parallel Process. Lett.* 21, 173–193, 2011.

[9] Varbanescu, A. L., Hijma, P., Van Nieuwpoort, R., and Bal, H., "Towards an effective unified programming model for many-cores." In *IPDPS*, pp. 681–692. IEEE, 2011.

[10] OpenACC. *Directives for Accelerators*. http://www.openacc-standard.org, 2017.

[11] Robison, A. D., Cilk plus: Language support for thread and vector parallelism. *Talk at HP-CAST*, 18:25, 2012.

[12] Leiserson, C. E., "The cilk++ concurrency platform." In *Design Automation Conference, 2009. DAC'09. 46th ACM/IEEE*, pp. 522–527. IEEE, 2009.

[13] Saule, E., and Çatalyürek, Ü. V., "An early evaluation of the scalability of graph algorithms on the intel mic architecture." In *Parallel and Distributed Processing Symposium Workshops and PhD Forum (IPDPSW), 2012 IEEE 26th International*, pp. 1629–1639. IEEE, 2012.

[14] De Dinechin, B. D., Van Amstel, D., Poulhiés, M., and Lager, G., "Time-critical computing on a single-chip massively parallel processor." In *DATE*, 2014.

[15] CEA STMicroelectronics. Platform 2012: A many-core programmable accelerator for ultra-efficient embedded computing in nanometer technology. *Whitepaper,* 2010.

[16] Texas Instruments. *SPRS866: 66AK2H12/06 Multicore DSP+ARM Key-Stone II System-on-Chip (SoC).*

[17] Kegel, P., Schellmann, M., and Gorlatch, S., "*Using OpenMP vs. Threading Building Blocks for Medical Imaging on Multi-Cores.*" In *Europar*. Springer, 2009.

[18] Lee, S., Min, S-J., and Eigenmann, R., OpenMP to GPGPU: A Compiler Framework for Automatic Translation and Optimization. *SIGPLAN Not.* 44, 101–110, 2009.

[19] Shen, J., Fang, J., Sips, H., and Varbanescu, A. L., "Performance gaps between OpenMP and OpenCL for multi-core CPUs." In *ICPPW*, pp. 116–125. IEEE, 2012.

[20] Krawezik, G., and Cappello, F., "Performance comparison of MPI and three OpenMP programming styles on shared memory multiprocessors." In *SPAA*. ACM, 2003.

[21] Kuhn, B., Petersen, P., and O'Toole, E., OpenMP versus threading in C/C++. *Concurr. Pract. Exp.* 12, 1165–1176, 2000.

[22] *OpenMP 2.5 Application Programming Interface*. http://www.openmp. org/wp-content/uploads/spec25.pdf, 2005.

[23] *OpenMP 3.0 Application Programming Interface*. http://www.openmp. org/wp-content/uploads/spec30.pdf, 2008.

[24] *OpenMP 4.0 Application Programming Interface*. http://www.openmp. org/wp-content/uploads/OpenMP4.0.0.pdf, 2013.

[25] *OpenMP 4.5 Application Programming Interface*. http://www.openmp. org/wp-content/uploads/openmp-4.5.pdf, 2015.

[26] *OpenMP Technical Report 2 on the OMPT Interface.* http://www. openmp.org/wp-content/uploads/ompt-tr2.pdf, 2014.

[27] *OpenMP Technical Report 4: Version 5.0 Preview 1.* http://www. openmp.org/wp-content/uploads/openmp-tr4.pdf, 2016.

[28] *OpenMP Technical Report 5: Memory Management Support for OpenMP 5.0.* http://www.openmp.org/wp-content/uploads/openmp-TR5-final.pdf, 2017.

[29] Podobas, A., and Karlsson, S., "Towards Unifying OpenMP Under the Task-Parallel Paradigm." In *IWOMP*, 2016.

[30] Buttazzo, G. C., *Hard Real-Time Computing Systems: Predictable Scheduling Algorithms and Applications*, volume 24. Springer Science and Business Media, 2011.

[31] Davis, R. I., and Burns, A., A survey of hard real-time scheduling for multiprocessor systems. *ACM computing surveys (CSUR)*, 43, 35, 2011.

[32] Bonifaci, V., Marchetti-Spaccamela, A., Stiller, S., and Wiese, A., "Feasibility analysis in the sporadic dag task model." In *Real-Time Systems (ECRTS), 2013 25th Euromicro Conference on*, pp. 225–233. IEEE, 2013.

[33] Baruah, S., Bonifaci, V., Marchetti-Spaccamela, A., Stougie, L., and Wiese, A., "A generalized parallel task model for recurrent real-time processes." In *Real-Time Systems Symposium (RTSS), 2012 IEEE 33rd*, pp. 63–72. IEEE, 2012.

[34] Saifullah, A., Li, J., Agrawal, K., Lu, C., and Gill, C., Multi-core real-time scheduling for generalized parallel task models. *Real-Time Sys.* 49, 404–435, 2013.

[35] Baruah, S., "Improved multiprocessor global schedulability analysis of sporadic dag task systems." In *Real-Time Systems (ECRTS), 2014 26th Euromicro Conference on*, pp. 97–105. IEEE, 2014.

[36] Li, J., Agrawal, K., Lu, C., and Gill, C., "Outstanding paper award: Analysis of global edf for parallel tasks." In *Real-Time Systems (ECRTS), 2013 25th Euromicro Conference on*, pp. 3–13. IEEE, 2013.

[37] Radojković, P., Girbal, S., Grasset, A., Quiones, E., Yehia, S., and Cazorla, F. J., On the evaluation of the impact of shared resources in multithreaded cots processors in time-critical environments. *ACM Trans. Architec. Code Opt. (TACO)*, 8:34, 2012.

[38] Rochange, C., Bonenfant, A., Sainrat, P., Gerdes, M., Wolf, J., Ungerer, T., et al. "Wcet analysis of a parallel 3d multigrid solver executed on the merasa multi-core." In *OASIcs-OpenAccess Series in Informatics*, volume 15. Schloss Dagstuhl-Leibniz-Zentrum fuer Informatik, 2010.

[39] Royuela, S., Ferrer, R., Caballero, D., and Martorell, X., "Compiler analysis for openmp tasks correctness." In *Proceedings of the 12th ACM International Conference on Computing Frontiers*, p. 7. ACM, 2015.

[40] Muchnick, S. S., *Advanced Compiler Design Implementation*. Morgan Kaufmann, 1997.

[41] Aho, A. V., Sethi, R., and Ullman, J. D., *Compilers: Principles, Techniques, and Tools*, volume 2. Addison-wesley Reading, 2007.

[42] Pereira, F. M. Q., Rodrigues, R. E., and Campos, V. H. S., "A fast and low-overhead technique to secure programs against integer overflows."

In *Proceedings of the 2013 IEEE/ACM International Symposium on Code Generation and Optimization (CGO)*, pp 1–11. IEEE Computer Society, 2013.

[43] Wilhelm, R., Engblom, J., Ermedahl, A., Holsti, N., Thesing, S., Whalley, D., et al. The worst-case execution-time problem?"overview of methods and survey of tools. *ACM Trans. Embed. Comput. Sys. (TECS)*, 7:36, 2008.

[44] McCabe, T. J., "A complexity measure." *IEEE Transactions on software Engineering*, 4, pp. 308–320, 1976.

4

Mapping, Scheduling, and Schedulability Analysis

Paolo Burgio[1], Marko Bertogna[1], Alessandra Melani[1], Eduardo Quiñones[2] and Maria A. Serrano[2]

[1]University of Modena and Reggio Emilia, Italy
[2]Barcelona Supercomputing Center (BSC), Spain

This chapter presents how the P-SOCRATES framework addresses the issue of scheduling multiple real-time tasks (RT tasks), made of multiple and concurrent non-preemptable *task parts*. In its most generic form, the scheduling problem in the architectural framework is a dual problem: scheduling task-to-threads, and scheduling thread-to-core replication.

4.1 Introduction

In our framework, we assume threads in the same OpenMP application are statically *pinned* to the available cores in the platforms[1]. This approach has two advantages: (i) the lower layer of the software stack, namely the runtime and the operating system (OS) support, are much simpler to design and implement; and (ii) we remove one dimension from the scheduling problem, that is, we only need to solve the problem of assigning tasks (in our case, OpenMP task parts) to threads/cores. For this reason, and limited to this chapter, we use the words "mapping" and "scheduling" interchangeably. As explained in Chapter 3, when a task encounters a task scheduling point (TSP), program execution branches into the OpenMP runtime, where task-to-thread mapping can: (1) begin the execution of a task region bound to the current team or (2) resume any previously suspended task region bound to the current

---

[1]Still, to enable multitasking at the OS level, the OS can preempt threads from one OpenMP application in favour of another OpenMP application.

team, as defined by the *parallel* OpenMP construct. Note that, the order in which these two actions are applied is not specified by the standard. An ideal task scheduler will schedule tasks for execution in a way that maximizes concurrency while accounting for load imbalance and locality to facilitate better performance.

The following part of the chapter describes the design of a simple partitioned scheduler, detailing how to enforce a limited-preemption scheduling policy to limit the overhead related to context switches whenever higher-priority instances arrive while the cores are busy executing lower-priority workload. It is also called *static* approach.

Then, we introduce the so-called dynamic approach, where scheduling happens with the adoption of a global queue where all tasks are inserted, and from where they can potentially be fetched by any worker in the system. We also show how it can be enhanced to support task migration across computing threads and cores, in a work-conservative environment.

In the following part, we describe our overall framework for the schedulability analysis, and then we specialize it for static/partitioned approach and dynamic/global approach, respectively.

We then briefly discuss the scheduling problem in the multi-core system that powers the four I/O clusters present in the fabric.

4.2 System Model

In the framework, an application may consist of multiple RT task instances, each one characterized by a different period or minimum inter-arrival time, deadline and execution requirement (see Figure 4.1). Each RT task starts executing on the host processor and may include (OpenMP-compliant) parallel workloads to be offloaded to the many-core accelerator. Such a parallel workload needs then to be scheduled on the available processing elements (PEs).

The parallel execution of each RT task is represented by a direct acyclic graph (DAG) composed of a set of nodes representing task parts. Nodes are connected through edges that represent precedence constraints among different task parts of the same offload. A task part can be executed only if all nodes that have a precedence constraint over it have already been executed.

To comply with the OpenMP semantics, an RT task is not directly scheduled on the PEs. Instead, its parallel workload is first mapped to several OS threads (up to the number of PEs available), and then these OS threads are scheduled onto the available cores. Figure 4.2 summarizes

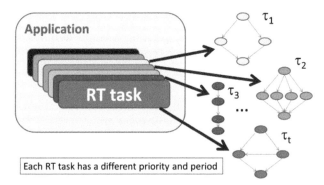

Figure 4.1 An application is composed of multiple real-time tasks.

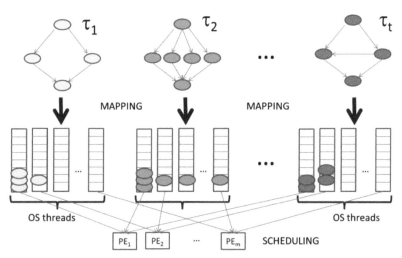

Figure 4.2 RT tasks are mapped to OS threads, which are scheduled on the processing elements.

the mapping/scheduling framework. Here, only partitioned/static approach is shown, where there is one task queue for each worker thread. In a fully dynamic/global approach, there is only one queue for every RT task, where all threads push and fetch work.

The number of OS threads onto which an RT task is mapped depends on mapping decisions. If the RT task does not present a large parallelism, it makes no sense to map it onto more than a limited number of threads. If instead the RT task has massively parallel regions, it may be useful to map it to a higher number of threads, up to the number of PEs in the many-core accelerator. A correct decision should consider the trade-off between the OS overhead implied by many threads and the speed-up obtainable with a larger run-time parallelism. Note that creating a larger number of threads than necessary may impose a significant burden to the OS, which needs to maintain the context of these threads, schedule, suspend, and resume them, with an obvious increase in the system overhead.

The architectural template targeted in the project (described in Chapter 2) is a many-core platform where cores are grouped onto clusters. The testbed accelerator, Kalray MPPA of the "Bostan" generation, has 256 cores grouped into 16 clusters of 16 cores each. We consider only the threads offloaded to the same cluster. Note that the intra-cluster scheduling problem is the main problem to solve in our scheduling framework. The reason is that P-SOCRATES adopts an execution model, where the context of each RT task that may need to be accelerated is statically offloaded to the target clusters before runtime.

For the above reasons, the main problem is therefore how to efficiently activate and schedule the threads associated with the different RT tasks that have been offloaded to the same cluster. The threads of each RT task will contend to execute on the available PEs with the threads of the other RT tasks. A smart scheduler will therefore need to decide which thread, or set of threads, to execute at any time in each PE of the considered cluster, such that all scheduling constraints are met. Depending on the characteristics of the running RT tasks (priority, period, deadline, etc.) the scheduler may choose to preempt an executing thread or set of threads, to schedule a different set of threads belonging to a higher priority (or more urgent) RT task.

4.3 Partitioned Scheduler

In a traditional partitioned scheduler, OS threads are statically assigned to cores, so that no thread may migrate from one core to another. The scheduling problem then reduces to the design of a single-core scheduler. We start from this approach and design our task-to-thread scheduler.

4.3.1 The Optimality of EDF on Preemptive Uniprocessors

The earliest deadline first (EDF) scheduling algorithm assigns scheduling priority to jobs according to their absolute deadlines: the earlier the deadline, the greater the priority (with ties broken arbitrarily). EDF is known to be *optimal* for scheduling a collection of jobs upon a preemptive uniprocessor platform, in the sense that *if a given collection of jobs can be scheduled to meet all deadlines, then the EDF-generated schedule for this collection of jobs will also meet all deadlines* [1]. To show that a system is EDF-schedulable upon a preemptive uniprocessor, it suffices to show the existence of a schedule meeting all deadlines — the optimality of EDF ensures that it will find such a schedule. Unfortunately, most of the commercial RTOSes do not implement the EDF scheduling policy. The main reasons are found in the added complexity of the scheduler, requiring timers to keep track of the thread deadlines, and in the agnostic behavior with respect to higher-priority work-load. This last concern is particularly important for industrial applications that have a set of higher-priority instances whose execution cannot be delayed. With an EDF scheduler, a lower-priority instance overrunning its expected budget may end up causing a deadline miss of a higher priority instance that has a later deadline. Instead, with a Fixed Priority (FP) scheduler, higher-priority jobs are protected against lower-priority overruns, because they will always be able to preempt a misbehaving lower-priority instance. This makes FP scheduling more robust for mixed-criticality scenarios where RT tasks of different criticality may contend for the same PEs. For the importance of FP scheduling, we decided to implement a partitioned scheduler based on this policy.

4.3.2 FP-scheduling Algorithms

In an FP-scheduling algorithm, each thread is assigned a distinct priority (as in P-SOCRATES scheduling model) and every instance (a.k.a. job/RT task instance) released by the thread inherits the priority of the associated thread.

The rate-monotonic (RM) scheduling algorithm [1] is an FP-scheduling algorithm in which the priorities of the tasks are defined based on their period: tasks with a smaller period are assigned greater priority (with ties broken arbitrarily). It is known [1] that RM is an optimal FP-scheduling algorithm for scheduling threads with relative deadlines equal to their minimum inter-arrival times upon preemptive uniprocessors: if there is any FP-scheduling algorithm that can schedule a given set of implicit-deadline threads to always meet all deadlines of all jobs, then RM will also always meet all deadlines of all jobs.

The deadline monotonic (DM) scheduling algorithm [2] is another FP-scheduling algorithm in which the priority of a task is defined based on its relative deadline parameter rather than its period: threads with smaller relative deadlines are assigned greater priority (with ties broken arbitrarily). Note that RM and DM are equivalent for implicit deadline systems, since all threads in such systems have their relative deadline parameters equal to their periods. It has been shown in [2] that DM is an optimal FP-scheduling algorithm for scheduling sets of constrained-deadline threads upon preemptive uniprocessors: if there is any FP-scheduling algorithm that can schedule a given constrained-deadline system to always meet all deadlines of all jobs, then DM will also always meet all deadlines of all jobs. DM is, however, known to not be optimal for systems where threads may have a deadline larger than their period.

4.3.3 Limited Preemption Scheduling

Preemption is a key concept in real-time scheduling, since it allows the OS to immediately allocate the processor to threads requiring urgent service. In fully preemptive systems, the running thread can be interrupted at any time by another thread with higher priority and be resumed to continue when all higher priority threads have completed. In other systems, preemption may be disabled for certain intervals of time during the execution of critical operations (e.g., interrupt service routines, critical sections, etc.). In other situations, preemption can be completely forbidden to avoid unpredictable interference among threads and achieve a higher degree of predictability (although higher blocking times).

The question of whether to enable or disable preemption during thread execution has been investigated by many authors under several points of view and it is not trivial to answer. A general disadvantage of the non-preemptive discipline is that it introduces additional blocking time in higher-priority threads, thereby reducing schedulability. On the other hand, preemptive scheduling may add a significant overhead due to context switches, significantly increasing the worst-case execution time. Both situations are schematized in Figure 4.3. CRPD in the figure stands for Cache-Related Preemption Delay, that is, the time overhead added to tasks' execution time due to cache cooling after a preemption.

There are several advantages to be considered when adopting a non-preemptive scheduler. Arbitrary preemptions can introduce a significant runtime overhead and may cause high fluctuations in thread-execution times, which degrades system predictability. Specifically, at least four different types of costs need to be considered at each preemption:

1. *Scheduling cost*: It is the time taken by the scheduling algorithm to suspend the running thread, insert it into the ready queue, switch the context, and dispatch the new incoming thread.

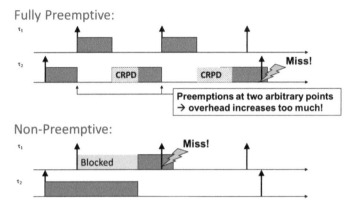

Figure 4.3 Fully preemptive vs. non-preemptive scheduling: preemption overhead and blocking delay may cause deadline misses.

2. *Pipeline cost*: It accounts for the time taken to flush the processor pipeline when the thread is interrupted, and the time taken to refill the pipeline when the thread is resumed.

3. *Cache-related cost*: It is the time taken to reload the cache lines evicted by the preempting thread. The WCET increment due to cache interference can be very large with respect to the WCET measured in non-preemptive mode.

4. *Bus-related cost*: It is the extra bus interference for accessing the next memory level due to the additional cache misses caused by preemption.

In order to predictably bound these penalties without sacrificing schedulability, we decided to adopt a limited preemption scheduler, which represents a trade-off between fully preemptive and non-preemptive scheduling. Note that this seamlessly integrates into the standard OpenMP tasking/execution model, where tasks can be preempted only at well-defined TSPs. See also Chapter 3.

4.3.4 Limited Preemption Schedulability Analysis

As in the fully preemptive case, the schedulability analysis of limited preemptive scheduling can be done analyzing the critical instant that leads to the worst-case response time of a given thread. However, differently from the fully preemptive case, the critical instant is not given by the synchronous arrival sequence, where all threads arrive at the same time, and all successive instances are released as soon as possible. Instead, in the presence of non-preemptive regions, the additional blocking from lower priority threads must be taken into account. Hence, the critical instant for a thread τ_i occurs when it is released synchronously and periodically with all higher priority threads, while the lower priority thread that is responsible of the largest blocking time of τ_i is released one unit of time before τ_i.

However, the largest response time of a thread is not necessarily due to the first job after a critical instant but might be due to later jobs. Therefore, as shown in [3], the schedulability analysis needs to check all τ_i's jobs within a given period of interest that goes from the above described critical instant until the first idle instant of τ_i. Let K_i be the number of such jobs.

When analyzing the schedulability of limited preemptive systems, a key role is played by the last non-preemptive region. Let q_i^{last} be the length of the last non-preemptive region of thread τ_i. When such a value is large, the response time of τ_i may decrease because the execution of many higher-priority instances is postponed after the end of τ_i, thus not interfering with τ_i. This allows improving the schedulability over the fully preemptive approach.

The blocking tolerance β_i of thread τ_i is defined as the maximum blocking that τ_i can tolerate without missing its deadline. Such a value may be computed by the following pseudo-polynomial relation:

$$\beta_i = \min_{k\in[1,K_i]} \max_{t\in\Pi_{i,k}} \left\{ t - kC_i + q_i^{last} - \sum_{j=1}^{i-1} \left\lceil \frac{t}{T_j} \right\rceil C_j \right\}$$

where $\Pi_{i,k}$ is the set of release times of jobs within the period of interest. The maximum allowed non-preemptive region of a τ_k is then given by:

$$NPR_k^{\max} \leftarrow \min_{i<k}\{\beta_i\}$$

Such a value determines the maximum spacing between two consecutive preemption points for each thread τ_k.

4.4 Global Scheduler with Migration Support

4.4.1 Migration-based Scheduler

The scheduling problem for single-core systems has already been solved with optimal priority assignments and scheduling algorithms back in the 1970s. In particular, RM assigning priorities with decreasing task periods, and DM assigning priorities with decreasing relative deadlines, are optimal priority assignments for sporadic systems with, respectively, implicit and constrained deadlines. This means that if a sporadic or synchronous periodic task system can be scheduled with fixed priorities on a single processor, then it can also be scheduled using RM (for implicit deadlines) [4] or DM (for constrained deadlines) [2]. Also, the EDF — that schedules at each time-instant the ready job with the earliest absolute deadline — is an optimal scheduling algorithm for scheduling arbitrary collections of jobs on a single processor [3, 4]. Therefore, if it is possible to schedule a set of jobs such that all deadlines are met, then the same collection of jobs can be successfully scheduled by EDF as well. These observations allowed us to optimally select the scheduling policies for the partitioned scheduler that we will describe shortly.

When allowing tasks to migrate among different cores, such as in the case of OpenMP untied task model (see Chapter 3 for further information), things are much more complicated: EDF, RM, and DM are no more optimal and can fail even at very low utilizations (arbitrarily close to one) due to the so-called Dhall's effect [5]. Still, these are unlucky corner cases which do not often recur in practice. The alternative approaches that allow higher schedulability ratios are dynamic algorithms that however lead to a higher number of preemptions and migrations, allowing the priority of a job to change multiple times. Examples are Pfair [6, 7], BF [8], LLREF [9], EKG [10], E-TNPA [11], LRE-TL [12], DP-fair [13], BF^2 [14, 15], and RUN [16]. The optimality of the above algorithms holds under very restrictive circumstances, i.e., neglecting preemption and migration overhead, and for sequential sporadic tasks with implicit deadlines. In this case, they are able to reach a full schedulable utilization, equal to the number of processors. Instead, they are not optimal when tasks may have deadlines different from periods (it has been shown in [17] that an optimal scheduler would require clairvoyance), for more general task models including parallel regions, limited preemptions, and/or DAG-structures, as with the task models adopted in the P-SOCRATES project.

The additional complexity inherent to the implementation, runtime overhead, scheduling and schedulability analysis of dynamic scheduling algorithms, as well as in the lack of optimality properties with relation to the task model adopted in the project, made their applicability to the considered setting questionable. For this reason, we decided to opt for the static priority class of scheduling algorithms, which is far more used in a practical setting due to some particularly desired features. Systems scheduled with static priority algorithms are rather easy to implement and to analyze; they allow reducing the response time of more critical tasks by increasing their priorities; they have a limited number of preemptions (and therefore migrations), bounded by the number of jobs activations in a given interval; they allow selectively refining the scheduling of the system by simply modifying the priority assignment, without needing to change the core of the scheduling algorithm (a much more critical component); they are easier to debug, simplifying the understanding of system monitoring traces and making it more intuitive to figure out why/when each task executes on which core; they are far more composable, i.e., changing any timing parameter of a lower-priority task does not alter the schedule of a higher-priority one, avoiding the need to recheck and re-validate the whole system.

4.4.2 Putting All Together

In our scheduling framework, the global scheduler will therefore consist of a fixed-priority scheduling algorithm. Each RT task is assigned a fixed priority, which is inherited by each one of its threads (there are at most m threads for each RT task). Threads that are *ready* to execute are ordered according to their priority in a global queue ("ready queue") from which the scheduler selects the m highest priority ones for execution, being m the number of available cores. These executing threads are popped from the queue and they change their state to *running*. New thread activations and incoming offloads are queued in the ready queue, based on their priorities. A blocked queue is also maintained with all suspended or waiting threads. Whenever a waiting thread is awakened, e.g., because the condition it was waiting for was satisfied, it is removed from the blocked queue and re-inserted into the ready queue according to its priority.

If the newly activated thread has a priority higher than one of the m *running* tasks, a preemption may take place, depending on the adopted preemption policy. With a fully preemptive scheduler, the preemption takes place immediately, as soon as the thread is (re-)activated. With a non-preemptive policy, the preemption is postponed until one of the running tasks finishes its execution. For this framework, we decided to adopt a limited preemption policy. According to this policy, threads are non-preemptively executed until they reach one of the statically defined preemption points, where they can be preempted if a higher priority thread is waiting to execute. This policy allows decreasing the preemption and migration overhead of fully preemptive policies, without imposing the excessive blocking delays experienced with non-preemptive approaches.

The problem with adopting a limited preemption scheduling policy is that it is necessary to define at which points to allow a preemption for each thread. Since requiring the programmer to manually insert suitable context-switch locations overly increases the programming complexity, we decided to automate the process by using meaningful information coming from the OpenMP mapping layer. In particular, the concept of TSP will be exposed to the scheduling layer in order to take informed decisions on when and where to allow a preemption. We will now detail this strategy.

4.4.3 Implementation of a Limited Preemption Scheduler

Arbitrary preemptions can introduce a significant runtime overhead and high fluctuations in thread-execution times, which degrades system predictability. These variations are due to multiple factors, including the time taken by the scheduling algorithm to suspend the running thread, insert it into the ready queue, switch the context, and dispatch the new incoming thread; the time taken to reload the cache lines evicted by the preempting thread; and the extra bus interference for accessing the next memory level due to the additional cache misses caused by preemption. Conversely, completely forbidding preemptions may cause an intolerable blocking to higher priority threads, potentially affecting their schedulability. For example, consider a system where a low-priority activity offloaded a parallel workload executing on all available cores. If a higher priority RT task now requests a subset of the cores to execute more important activities, it will need to wait until the low-priority ones are finished, eventually leading to a deadline miss. Such a miss could have been easily avoided by allowing preemptions.

With the limited preemption scheduling model adopted in the project, threads will execute non-preemptively until they reach a TSP. At these points, the execution control is moved back to the OpenMP runtime to decide which task (part) to map on that thread. Essentially, the mapper will fetch one of the tasks (belonging to the offload associated to the considered thread) that are ready to execute and map it to that thread. These are points that mark an interruption in the task-execution flow, potentially leading to context switches and/or some memory locality loss. In other words, TSPs are good candidate to be selected for potential preemption points, since they may represent a discontinuity in the continuous execution of a task, potentially requiring a new task to load new data to local memory. Taking advantage of these points seems reasonable to guarantee a reduced pollution of cache locality of an executing task, allowing a thread context switch only when a preemption causes less harm.

However, it remains to be shown how the information from the OpenMP runtime is to be propagated to the RTOS scheduling layer. Note that every RT task that is offloaded to the accelerator is managed by an instance of custom OpenMP runtime. This instance, among the other tasks, keeps track of the dependencies among the nodes of the RT task (the OpenMP task parts), and schedules for execution only those nodes whose dependencies have been satisfied. When a thread fetches a task from the pool for execution, it will continue uninterruptedly until it reaches a TSP. At this point, the runtime regains control, and it may decide to invoke the OS scheduler using a simple function call. The scheduler can then check whether there are new offload requests pending and/or there are blocked tasks that have been awakened. Potential higher-priority threads arrivals will then trigger a preemption, saving the context of the preempted thread and scheduling the higher priority one.

In this way, the OpenMP semantics of TSPs are propagated at RTOS scheduling level, allowing smarter decisions on the preemption locations. The timing analysis will also be significantly easier, since it will be sufficient to analyze the worst-case execution requirements of each task part, knowing that such code blocks will be executed without interruptions. The timing characterization of each task part will factor in the worst-case delay related to interfering instances, assuming each task part needs to (re-)load all required data from scratch. This makes the analysis robust and tractable, without requiring the timing analyzer to consider all possible instructions as potential preemption points but characterizing only the worst-case timing parameters of each individual task part. In Chapter 5, it is described how to obtain the maximum execution time of a task part, with and without including the additional time-penalty due to interference with other applications running concurrently. These two timing estimates are added to the characterization of every task part in the TDG produced by the compiler. This new TDG annotated with timing information is called the OpenMP-TDG and serves as input to our schedulability analysis.

Still, one may further reduce the number of potential preemption points, by not invoking the OS scheduler at each TSP. For example, with a Breadth-First mapping model, a task creating additional tasks will continue executing on its thread, without leading to a (task-level) context switch. In this case, it may be better not to invoke the OS scheduler at TSPs coinciding with task-creation directives, since the original task may continue executing without any discontinuity in the local context. A smarter option can be to invoke the OS scheduler only when the runtime decides to map a different task next (e.g., because the current one is finished, or due to a work-first strategy, or because of a taskwait directive). These TSPs are more likely to lead to a cache locality loss, reducing the additional impact due to preemptions.

That said, in order to simplify the schedulability analysis and avoid long non-preemptive regions, we decided to invoke the scheduler at each TSP. Although it may be beneficial to reduce the number of preemption points, we opted for the simplest solution that allows us to provide a proof of concept of the proposed approach. In the evaluation phase, we will then identify the impact of the preemption points to the scheduling overhead.

4.5 Overall Schedulability Analysis

We now will describe the overall schedulability analysis of systems executing within the P-SOCRATES framework. The analysis is based on the computation of the worst-case response time of RT tasks concurrently executing on a given cluster of cores. Two different analyses are presented, depending on the mapping/scheduling mechanisms supported by the framework: (i) a dynamic solution based on a global scheduler allowing a work-conserving behavior, and (ii) a fully static solution based on a partitioned scheduler and a fixed task-to-thread mapping.

4.5.1 Model Formalization

On our overall framework, an OpenMP program is composed of recurring instances of a RT task (identified with a *target* OpenMP construct), which in turn is composed of task parts. Without loss of generality, in this paragraph, we consider [18] a set $\tau = \{\tau_1, \ldots, \tau_n\}$ of n sporadic conditional parallel tasks (cp-tasks) that execute upon a platform consisting of m identical processors. Each cp-task τ_k releases a potentially infinite sequence of jobs. Each job of τ_k is separated from the next by at least T_k time-units and has a constrained relative deadline $D_k <= T_k$. Moreover, each cp-task τ_k is represented as a directed acyclic graph $G_k = (V_k, E_k)$, where $V_k = \{v_{k,1}, \ldots, v_{k,nk}\}$ is a set of nodes (or vertices) and E_k is a set of directed arcs (or edges), as shown in Figure 4.4. Each node $v_{k,j}$ represents a sequential chunk of execution (or "sub-task") and is characterized by a worst-case execution time $C_{k,j}$. Preemption and migration overhead is assumed to be integrated within the WCET values, as given by the timing analysis. Arcs represent dependencies between sub-tasks, that is, an edge $(v_{k,1}, v_{k,2})$ means that $v_{k,1}$ must complete before $v_{k,2}$ can start executing. A node with no incoming arcs is referred to as

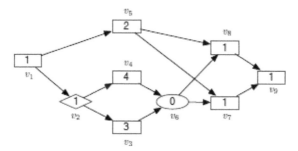

Figure 4.4 A sample cp-task. Each vertex is labeled with the WCET of the corresponding sub-task.

a source, while a node with no outgoing arcs is referred to as a sink. Without loss of generality, each cp-task is assumed to have exactly one source v_k^{source} and one sink node v_k^{sink}. If this is not the case, a dummy source/sink node with zero WCET can be added to the DAG, with arcs to/from all the source/sink nodes. The subscript k in the parameters associated with the task τ_k is omitted whenever the reference to the task is clear in the discussion.

In the cp-task model, nodes can be of two types:

1. Regular nodes, represented as rectangles, allow all successor nodes to be executed in parallel;
2. Conditional nodes, coming in pairs and denoted by diamonds and circles, represent the beginning and the end of a conditional construct, respectively, and require the execution of exactly one node among the successors of the start node.

Please note that this is a general solution for scheduling parallel recurring RT-Dags. In the specific domain of this project, where OpenMP is used as a frontend to specify DAGS, it may occur that the compiler cannot fully extract the DAG because there are conditionals that cannot be statically solved. See Section 3.4.3.2, "Missing information of the DAG", in Chapter 3, for a discussion about this issue.

To properly model the possible execution flows, a further restriction is imposed to the connections within a conditional branch. That is, a node belonging to a branch of a conditional statement cannot be connected to nodes outside that branch (including other branches of the same statement). This is formally stated in the following definition.

Definition 4.1. Let (v_1, v_2) be a pair of conditional nodes in a DAG $G_k = (V_k, E_k)$. The pair (v_1, v_2) is a conditional pair if the following holds:

1. If there are exactly q outgoing arcs from v_1 to nodes $s_1, s_2, \ldots, s_q$, for some $q > 1$, then there are exactly q incoming arcs into v_2 in E_k, from some nodes $t_1, t_2, \ldots, t_q$.
2. For each $1 \in \{1, 2, \ldots, q\}$, let $V_l'E_l'$ denote all the nodes and arcs on paths reachable from sl that do not include node v_2. By definition, s_l is the sole source node of the DAG $Gl':= (V_l'E_l')$. It must hold that tl is the sole sink node of G_l'.
3. It must hold that V_l' and V_j' have a null intersection, for all $1 \neq j$. Additionally, with the exception of (v_1, s_l) there should be no arcs in E_k into nodes in V_l' from nodes not in V_l', for each l in $\{1, 2, \ldots, q\}$.

A chain or path of a cp-task τ_k is a sequence of nodes $\lambda = (v_{k,a}, \ldots, v_{k,b})$ such that $(v_{k,j}, v_{k,j+1}) \in E_k$, for all $j \in [a,b]$. The length of a chain of τ_k, denoted by $len(\lambda)$, is the sum of the WCETs of all its nodes. The longest path of a cp-task is any source-sink path of the task that achieves the longest length.

Definition 4.2. The length of a cp-task τ_k, denoted by L_k, is the length of any longest path of τ_k.

Note that L_k also represents the minimum worst-case execution time of cp-task τ_k, that is, the time required to execute it when the number of processing units is sufficiently large (potentially infinite) to allow the task to always execute with maximum parallelism. A necessary condition for the feasibility of a cp-task τ_k is that $L_k \leq D_k$.

In the absence of conditional branches, the classical sporadic DAG task model defines the volume of the task as the worst-case execution time needed to complete it on a dedicated single-core platform. This quantity can be computed as the sum of the WCETs of all the sub-tasks, that is $\sum_{v_{k,j} \in V_k} C_{k,j}$. In the presence of conditional branches, assuming that all sub-tasks are always executed is overly pessimistic. Hence, the concept of volume of a cp-task is generalized by introducing the notion of worst-case workload.

Definition 4.3. The worst-case workload W_k of a cp-task τ_k is the maximum time needed to execute an instance of τ_k on a dedicated single-core platform, where the maximum is taken among all possible choices of conditional branches.

Section 4.5 will explain in detail how the worst-case workload of a task can be computed efficiently.

The utilization U_k of a cp-task τ_k is the ratio between its worst-case workload and its period, that is, $U_k = W_k/T_k$. For the task-set τ, its total utilization U is defined as the sum of the utilizations of all tasks. A simple necessary condition for feasibility is $U \leq m$.

Figure 4.4 illustrates a sample cp-task consisting of nine sub-tasks (nodes) $V = \{v_1,\ldots,v_9\}$ and 12 precedence constraints (arcs). The number inside each node represents its WCET. Two of the nodes, v_2 and v_6, form a conditional pair, meaning that only one sub-task between v_3 and v_4 will be executed (but never both), depending on a conditional clause. The length (longest path) of this cp-task is $L = 8$, and is given by the chain (v_1, v_2, v_4, v_6, v_7, v_9). Its volume is 14 units, while its worst-case workload must take into account that either v_3 or v_4 are executed at every task instance. Since v_4 corresponds to the branch with the largest workload, $W = 11$.

To further clarify the restrictions imposed to the graph structure, note that v_4 cannot be connected to v_5, because this would violate the correctness of conditional constructs and the semantics of the precedence relation. In fact, if they were connected and v_3 were executed, then v_5 would wait forever, since v_4 is not executed. For the same reason, no connection is possible between v_4 and v_3, as they belong to different branches of the same conditional statement.

In the following sections, we will consider the dynamic approach consisting of a best-effort mapper, coupled with a fixed priority global scheduler. RT tasks are indexed according to their priorities, being τ_1 the highest priority one. For details on the scheduling algorithm and mapping, please refer to P-SOCRATES project's Deliverable D3.3.2 [19]. To understand the following analysis, it is sufficient to observe that the adopted scheduler allows a work-conserving behavior, never idling a core whenever there is some pending workload to execute.

4.5.2 Critical Interference of cp-tasks

We now present a schedulability analysis for cp-tasks globally scheduled by any work-conserving scheduler. The analysis is based on the notion of *interference*. In the existing literature for globally scheduled sequential task systems, the interference on a task τ_k is defined as the sum of all intervals in which τ_k is ready, but cannot execute because all cores are busy executing other tasks. We modify this definition to adapt it to the parallel nature of cp-tasks, by introducing the concept of critical interference.

Given a set of cp-tasks τ and a work-conserving scheduler, we define the *critical chain* of a task as follows.

Definition 4.4. The critical chain λ_k^* of a cp-task τ_k is the chain of nodes of τ_k that leads to its worst-case response-time R_k.

The critical chain of cp-task τ_k is in principle determined by taking the sink vertex v_k^{sink} of the worst-case instance of τ_k (i.e., the job of τ_k that has the largest response-time in the worst-case scenario), and recursively pre-pending the last to complete among the predecessor nodes (whether conditional or not), until the source vertex $v_{k,1}$ has been included in the chain.

A critical node of task τ_k is a node that belongs to τ_k's critical chain. Since the response-time of a cp-task is given by the response-time of the sink vertex of the task, the sink node is always a critical node. For deriving the worst-case response-time of a task, it is then sufficient to characterize the maximum interference suffered by its critical chain.

Definition 4.5. The critical interference I_k on task τ_k is defined as the cumulative time during which some critical nodes of the worst-case instance of τ_k are ready, but do not execute because all cores are busy.

Lemma 4.1. Given a set of cp-tasks τ scheduled by any work-conserving algorithm on m identical processors, the worst-case response-time of each task τ_k is

$$R_k = \text{len}(\lambda_k^*) + I_k. \tag{4.1}$$

Proof. Let r_k be the release time of the worst-case instance of τ_k. In the scheduling window $[r_k, r_k + R_k]$, the critical chain will require $\text{len}(\lambda_k^*)$ time-units to complete. By Definition 4.5, at any time in this window in which τ_k does not suffer critical interference, some node of the critical chain is executing. Therefore $R_k - I_k = \text{len}(\lambda_k^*)$.

The difficulty in using Lemma 4.1 for schedulability analysis is that the term I_k may not be easy to compute. An established solution is to express the total interfering workload as a function of individual contributions of the interfering tasks, and then upper-bound such contributions with the worst-case workload of each interfering task τ_k.

In the following, we explain how such interfering contributions can be computed, and how they relate to each other to determine the total interfering workload.

Definition 4.6. The critical interference $I_{i,k}$ imposed by task τ_i on task τ_k is defined as the cumulative workload executed by sub-tasks of τ_k while a critical node of the worst-case instance of τ_k is ready to execute but is not executing.

Lemma 4.2. For any work-conserving algorithm, the following relation holds:

$$I_k = \frac{1}{m} \sum_{\tau_i \in \mathcal{T}} I_{i,k}. \tag{4.2}$$

Proof. By the work-conserving property of the scheduling algorithm, whenever a critical node of τ_k is interfered, all m cores are busy executing other sub-tasks. The total amount of workload executed by sub-tasks interfering with the critical chain of τ_k is then mI_k. Hence,

$$\sum_{\tau_i \in \mathcal{T}} I_{i,k} = mI_k.$$

By reordering the terms, the lemma follows.

Note that when $i = k$, the critical interference $I_{k,k}$ may include the interfering contributions of non-critical subtasks of τ_k on itself, that is, the self-interference of τ_k. By combining Equations (4.1) and (4.2), the response-time of a task τ_k can be rewritten as:

$$R_k = \text{len}(\lambda_k^*) + \frac{1}{m}I_{k,k} + \frac{1}{m}\sum_{\tau_i \in \mathcal{T}, i \neq k} I_{i,k}. \tag{4.3}$$

In the following, we will show how to provide upper bounds on the unknown terms of Equation (4.3) for systems adopting a global fixed-priority scheduler with preemption support.

4.5.3 Response Time Analysis

In this section, we derive an upper-bound on the worst-case response-time of each cp-task using Equation (4.3). To this aim we need to compute the interfering contributions $I_{i,k}$. In the sequel, we first consider the inter-task interference $(i \neq k)$ and then the intra-task interference $(i = k)$.

4.5.3.1 Inter-task interference

We divide the contribution to the workload of an interfering task τ_I in a window of interest between carry-in, body, and carry-out jobs. The *carry-in job* is the first instance of τ_i that is part of the window of interest and has release time before and deadline within the window of interest. The *carry-out job* is the last instance of τ_i executing in the window of interest, having a deadline after the window of interest. All other instances of τ_i are named *body jobs*. For sequential task-sets, an upper-bound on the workload of an interfering task τ_i within a window of length L occurs when the first job of τ_i starts executing as late as possible (with a starting time aligned with the beginning of the window of interest) and later jobs are executed as soon as possible (see Figure 4.5).

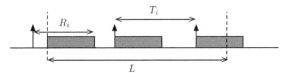

Figure 4.5   Worst-case scenario to maximize the workload of an interfering task τ_i in the sequential case.

For cp-task systems, it is more difficult to determine a configuration that maximizes the carry-in and carry-out contributions. In fact:

1. Due to the precedence constraints and different degree of parallelism of the various execution paths of a cp-task, it may happen that a larger workload is executed within the window if the interfering task is shifted left, i.e., by decreasing the carry-in and increasing the carry-out contributions. This happens for example when the first part of the carry-in job has little parallelism, while the carry-out part at the end of the window contains multiple parallel sub-tasks.
2. A sustainable schedulability analysis [10] must guarantee that all tasks meet their deadlines even when some of them execute less than the worst-case. For example, one of the sub-tasks of an execution path of a cp-task may execute for less than its WCET $C_{i,j}$. This may lead to larger interfering contributions within the window of interest (e.g., a parallel section of a carry-out job is included in the window due to an earlier completion of a preceding sequential section).
3. The carry-in and carry-out contribution of a cp-task may correspond to different conditional paths of the same task, with different levels of parallelism.

To circumvent the above issues, we consider a scenario in which each interfering job of task τ_i executes for its worst-case workload W_i, i.e., the maximum amount of workload that can be generated by a single instance of a cp-task. We defer the computation of W_i to Section 4.5.3. The next lemma provides a safe upper-bound on the workload of a task τ_i within a window of interest of length L.

Lemma 4.3. An upper-bound on the workloads of an interfering task τ_i in a window of

$$
\mathcal{W}_i(L) = \left\lfloor \frac{L + R_i - W_i/m}{T_i} \right\rfloor W_i \\
+ \min(W_i, m \cdot ((L + R_i - W_i/m) \bmod T_i)).
$$

length L is given by

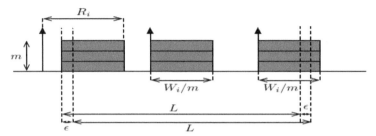

Figure 4.6   Worst-case scenario to maximize the workload of an interfering cp-task τ_i.

Proof. Consider a situation in which all instances of i execute for their worst-case workload W_i. The highest workload within a window of length L for such a task configuration is produced when the carry-in and carry-out contributions are evenly distributed among all cores, as shown in Figure 4.6. Note that distributing the carry-in or carry-out contributions on a smaller number of cores may not increase the workload within the window. Moreover, other task configurations with a smaller workload for the carry-in or carry-out instance cannot lead to a higher workload in the window of interest: although a reduced carry-in workload may allow including a larger part of the carry-out (as in shifting right the window of interest by $W_i = m$ in the figure), the carry-out part that enters the window from the right cannot be larger than the carry-in reduction.

An upper-bound on the number of carry-in and body instances that may execute within the window is

$$\left\lfloor \frac{L + R_i - W_i/m}{T_i} \right\rfloor,$$

each one contributing for W_i. The portion of the carry-out job included in the window of interest is $(L + R_i - W_i/m) \bmod T_i$. Since at most m cores may be occupied by the carryout job within that interval, and the carry-out job cannot execute for more than W_i units, the lemma follows.

4.5.3.2 Intra-task interference
We now consider the remaining terms of Equation (4.3), which take into account the contribution of the considered task to its overall response-time, and we compute an upper-bound on

$$Z_k \stackrel{\text{def}}{=} \operatorname{len}(\lambda_k^*) + \frac{1}{m} I_{k,k}.$$

Lemma 4.4. For a constrained deadline cp-task system scheduled with any work-conserving algorithm, the following relation holds for any task τ_k:

$$Z_k = \text{len}(\lambda_k^*) + \frac{1}{m} I_{k,k} \le L_k + \frac{1}{m}(W_k - L_k). \tag{4.4}$$

Proof. Since we are in a constrained deadline setting, a job will never be interfered with by other jobs of the same task. W_k being the maximum possible workload produced by a job of cp-task τ_k, the portion that may interfere with the critical chain λ_k is $W_k - \text{len}(\lambda_k^*)$. Then, $I_{k,k} \le W_k - \text{len}(\lambda_k^*)$. Hence,

$$\text{len}(\lambda_k^*) + \frac{1}{m} I_{k,k} \le \text{len}(\lambda_k^*) + \frac{1}{m}(W_k - \text{len}(\lambda_k^*)). \tag{4.5}$$

Since $\text{len}(\lambda_k^*) \le L_k$ and $m \ge 1$, the lemma follows.

Since Z_k includes only the contribution of task τ_k, one may think that the sum $[\text{len}(\lambda_k^*) + 1/m\, I_{k,k}]$ is equal to the worst-case response-time of τ_k when it is executed in isolation on the multi-core system (i.e., the makespan of τ_k).

However, this is not true. For example, consider the case of a cp-task τ_k with only one if-then-else statement; assume that when the "if" part is executed, the task executes one sub-task of length 10; otherwise, the task executes two parallel sub-tasks of length 6 each. When τ_k is executed in isolation on a two-core platform, the makespan is clearly given by the "if" branch, i.e., 10. When instead τ_k can be interfered with by one job of a task τ_i which executes a single sub-task of length 6, the worst-case response time of τ_k occurs when the "else" branch is executed, yielding a response time of 12. The share of the response time due to the term $\text{len}(\lambda_k^*) + 1/m\, I_{k,k}$ in Equation (4.3) is $6 + (1 = 2)6 = 9$, which is strictly smaller than the makespan. Note that $\text{len}(\lambda_k^*) + 1/m\, I_{k,k}$ does not even represent a valid lower bound on the makespan. This can be seen by replacing the "if" branch in the above example with a shorter subtask of length 8, giving a makespan of 8. For this reason, one cannot replace the term $\text{len}(\lambda_k^*) + 1/m\, I_{k,k}$ in Equation (4.4) with the makespan of τ_k.

The right-hand side of Equation (4.4) ($L_k + 1/m(W_k - L_k)$) has been therefore introduced to upper-bound the term $\text{len}(\lambda_k^*) + 1/m\, I_{k,k}$. Interestingly, this quantity does also represent a valid upper-bound on the makespan of τ_k, so that it can be used to bound the response time of a cp-task executing in isolation. We omit the proof that is identical to the proofs of the given bounds, considering only the interference due to the task itself.

4.5.3.3 Computation of cp-task parameters

The upper-bounds on the interference given by Lemmas 4.3, 4.4, and 4.5 require the computation of two characteristic parameters for each cp-task τ_k: the worst-case workload W_k and the length of the longest chain L_k. The longest path of a cp-task can be computed in exactly the same way as the longest path of a classical DAG task, since any conditional branch defines a set of possible paths in the graph. For this purpose, conditional nodes can be considered as if they were simply regular nodes. The computation can be implemented time linearly in the size of the DAG by standard techniques, see e.g., Bonifaci et al. [11] and references therein.

The computation of the worst-case workload of a cp-task is more involved. We hereafter show an algorithm to compute W_k for each task τ_k in time quadratic in the DAG size, whose pseudocode is shown in Algorithm 4.1.

The algorithm first computes a topological order of the DAG[2]. Then, exploiting the (reverse) topological order, a simple dynamic program can compute for each node the accumulated workload corresponding to the portion of the graph already examined. The algorithm must distinguish the case when the node under analysis is the head of a conditional pair or not.

Algorithm 4.1 Worst-case Workload Computation

1:	**procedure** WCW(G)
2:	$\sigma \leftarrow$ TOPOLOGICALORDER(G)
3:	$S(v^{\text{sink}}) \leftarrow \{v^{\text{sink}}\}$
4:	**for** $v_i \in \sigma$ from sink to source **do**
5:	**if** SUCC(v_i) $\neq \emptyset$ **then**
6:	**if** ISBEGINCOND(v_i) **then**
7:	$v^* \leftarrow \text{argmax}_{v \in \text{SUCC}(v_i)} C(S(v))$
8:	$S(v_i) \leftarrow \{v_i\} \cup S(v^*)$
9:	**else**
10:	$S(v_i) \leftarrow \{v_i\} \cup \bigcup_{v \in \text{SUCC}(v_i)} S(v)$
11:	**end if**
12:	**end if**
13:	**end for**
14:	**return** $C(S(v^{\text{source}}))$
15:	**end procedure**

---

[2]A topological order is such that if there is an arc from u to v in the DAG, then u appears before v in the topological order. A topological order can be easily computed in time linear in the size of the DAG (see any basic algorithm textbook, such as [17]).

If this is the case, then the maximum accumulated workload among the successors is selected; otherwise, the sum of the workload contributions of all successors is computed.

Algorithm 4.1 takes as input the graph representation of a cp-task G and outputs its worst-case workload W. In the algorithm, for any set of nodes S, its total WCET is denoted by C(S). First, at line 2, a topological sorting of the vertices is computed and stored in the permutation. Then, the permutation is scanned in reverse order, that is, from the (unique) sink to the (unique) source of the DAG. At each iteration of the for loop at line 4, a node v_i is analyzed; a set variable $S(v_i)$ is used to store the set of nodes achieving the worst-case workload of the subgraph including v_i and all its descendants in the DAG. Since the sink node has no successors, $S(v^{sink})$ is initialized to $\{v^{sink}\}$ at line 3. Then, the function SUCC(v_i) computes the set of successors of v_i. If that set is not empty, function ISBEGINCOND(v_i) is invoked to determine whether v_i is the head node of a conditional pair. If so, the node v^* achieving the largest value of C(S(v)), among v in SUCC(v_i), is computed (line 7). The set $S(v^*)$ therefore achieves the maximum cumulative worst-case workload among the successors of v_i, and is then used to create $S(v_i)$ together with v_i. Instead, whenever v_i is not the head of a conditional pair, all its successors are executed at runtime. Therefore, the workload contributions of all its successors must be merged into $S(v_i)$ (line 10) together with v_i. The procedure returns the worst-case workload accumulated by the source vertex, that is $C(S(v^{source}))$.

The complexity of the algorithm is quadratic in the size of the input DAG. Indeed, there are $O(|E|)$ set operations performed throughout the algorithm, and some operations on a set S (namely, the ones at line 7) also require computing C(S), which has cost $O(|V|)$. So, the time complexity is $O(|V|\,|E|)$. To implement the set operations, set membership arrays are sufficient.

One may be tempted to simplify the procedure by avoiding the use of set operations, keeping track only of the cumulative worst-case workload at each node, and allowing a linear complexity in the DAG size. However, such an approach would lead to an overly pessimistic result. Consider a simple graph with a source node forking into multiple parallel branches which then converge on a common sink. The cumulative worst-case workload of each parallel path includes the contribution of the sink. If we simply sum such contributions to derive the cumulative worst-case workload of the source, the contribution of the sink would be counted multiple times. Set operations are therefore needed to avoid accounting multiple times each node contribution.

We now present refinements of Algorithm 4.1 in special sub-cases of interest.

4.5.4 Non-conditional DAG Tasks

The basic sporadic DAG task model does not explicitly account for conditional branches. Therefore, all vertices of a cp-task contribute to the worst-case workload, which is then equal to the volume of the DAG task:

$$W_k = \sum_{v_{k,j} \in V_k} C_{k,j}.$$

In this particular case, the time complexity to derive the worst-case workload of a task (quadratic in the general case), becomes $O(|V|)$, i.e., linear in the number of vertices.

4.5.5 Series–Parallel Conditional DAG Tasks

Some programming languages yield series–parallel cp-tasks, that is, cp-tasks that can be obtained from a single edge by series composition and/or parallel composition. For example, the cp-task in Figure 4.5 is series–parallel, while the cp-tasks in Figures 4.2 and 4.6 are not. Such a structure can be detected in linear time [13]. In series–parallel graphs, for every head s_i of a conditional or parallel branch there is a corresponding tail t_i. For example, in Figure 4.5, the tail corresponding to parallel branch head v_2 is v_9. Algorithm 4.1 can be specialized to series–parallel graphs. For each vertex u, the algorithm will simply keep track of the worst-case workload of the subgraph reachable from u, as follows. For each head vertex s_i of a parallel branch, the contribution from all successors should be added to s_i's WCET, subtracting, however, the worst-case workload of the corresponding tail t_i a number of times equal to the out-degree of s_i minus 1; for each head vertex s_i of a conditional branch, only the maximum among the successors' worst-case workloads is added to s_i's WCET. Finally, for all non-head vertices add the worst-case workload of their unique successor to their WCET. The complexity of this algorithm reduces then to $O(|E|)$, i.e., it becomes linear in the size of the graph.

4.5.6 Schedulability Condition

Lemmas 4.3 and 4.4 and the bounds previously computed allow for proving the following theorem.

Theorem 4.1. Given a cp-task-set globally scheduled with global FP on m cores, an upper-bound R_k^{ub} on the response-time of a task τ_k can be derived by the fixed-point iteration of the following expression, starting with $R_k^{ub} = L_k$:

$$R_k^{ub} \leftarrow L_k + \frac{1}{m}(W_k - L_k) + \left\lfloor \frac{1}{m} \sum_{\forall i \neq k} \mathcal{X}_i^{ALG} \right\rfloor, \quad \mathcal{X}_i^{FP}$$

where:

$$\mathcal{X}_i^{FP} = \begin{cases} \mathcal{W}_i(R_k^{ub}), & \forall i < k \\ 0, & otherwise \end{cases} ;$$

because the interference from lower priority tasks can be neglected assuming a fully preemptive scheduler.

The schedulability of a cp-task system can then be simply checked using Theorem 4.1 to compute an upper-bound on the response-time of each task. In the FP case, the bounds are updated in decreasing priority order, starting from the highest priority task. In this case, it is sufficient to apply Theorem 4.1 only once for each task.

4.6 Specializing Analysis for Limited Pre-emption Global/Dynamic Approach

The response time analysis in Equation (4.3) can be easily extended [20] to incorporate the impact of the limited pre-emption strategy on DAG-based task-sets[3]. To do so, the factor that computes the inter-task interference must be augmented to incorporate the impact of lower-priority interference. Overall, the response time upper-bound can be computed as follows:

$$R_k^{ub} \leftarrow L_k + \frac{1}{m}(vol(G_k) - L_k) + \left\lfloor \frac{1}{m}(I_k^{l_p} + I_k^{h_p}) \right\rfloor$$

With LP, tasks are not only interfered with by higher-priority tasks, but also by already started lower-priority tasks whose execution has not reached a pre-emption point yet, and so cannot be suspended. In the worst-case scenario, when a high-priority task τ_k is released, all the m processors have just started executing the m largest NPRs of m different lower priority tasks. After τ_k started executing, it could be blocked again by at most $m - 1$ lower priority

---

[3]This section only considers LP with eager approach. In [28], we develop the analysis for Lazy approach as well. Interested readers are encouraged to refer to it for the complete analysis.

tasks at each pre-emption point. Therefore, for sequential task-sets, the lower priority interference is upper-bounded considering: (1) the set of the longest NPR of each lower-priority task and then (2) the sum of the m and $m-1$ longest NPRs of this set, as computed in [21]. This no longer holds for DAG-based task-sets, because multiple NPRs from the same task can execute in parallel. Next, we present two methods to compute the lower-priority interference in DAG-based task-sets.

4.6.1 Blocking Impact of the Largest NPRs (LP-max)

The easiest way of deriving the lower-priority interference is to account for the m and $m-1$ largest NPRs among all lower-priority tasks:

$$\Delta_k^m = \sum \max_{\tau_i \in lp(k)}^m \left(\max_{1 \le j \le q_{i+1}}^m C_{i,j} \right)$$

$$\Delta_k^{m-1} = \sum \max_{\tau_i \in lp(k)}^{m-1} \left(\max_{1 \le j \le q_{i+1}}^{m-1} C_{i,j} \right)$$

where $\sum max_{\tau_i \in lp(k)}^m$ and $\sum max_{\tau_i \in lp(k)}^{m-1}$ denote the sum of the m and $m-1$ largest values among the NPRs of all tasks $\tau_i \in$ lp(k) respectively, while $max_{1 \le j \le q_{i+1}}^m$ and $max_{1 \le j \le q_{i+1}}^{m+1}$ denote the m and $m-1$ largest NPRs of a task τ_i. Despite its simplicity, this strategy is pessimistic because it considers that the largest m and $m-1$ NPRs can execute in parallel, regardless of the precedence constraints defined in the DAG.

4.6.2 Blocking Impact of the Largest Parallel NPRs (LP-ILP)

The edges in the DAG determine the maximum level of parallelism a task may exploit on m cores, which in turn determines the amount of blocking impacting over higher-priority tasks. This information must therefore be incorporated in the analysis to better upper-bound the lower-priority interference. To do so, we propose a new analysis method that incorporates the precedence constraints among NPRs, as defined by the edges in the DAG, into the LP response-time analysis. Our analysis uses the following definitions:

Definition 4.7: The LP worst-case workload of a task executing on c cores is the sum of the WCET of the c largest NPRs that can execute in parallel.

Definition 4.8. The overall LP worst-case workload of a set of tasks executing on m cores is the maximum time used for executing this set in a given execution scenario, i.e. fixing the number of cores used for each task.

Given a task τ_k, our analysis derives the lower-priority interference of lp(k) by computing new Δ_k^m and Δ_k^{m+1} factors in a three-step process:

1. Identify the LP worst-case workload of each task in lp(k) when executing on 1 to m cores;
2. Compute the overall LP worst-case workload of lp(k) for all possible execution scenarios;
3. Select the scenario that maximizes the lower-priority interference.

In order to facilitate the explanation of the three steps, the next sections consider an lp(k) composed of four DAG-tasks $\{\tau_1, \tau_2, \tau_3, \tau_4\}$ (see Figure 4.7), executed on an $m = 4$ core platform.

The nodes (NPRs) of τ_i are labeled as $v_{i,j}$ with their WCET ($C_{i,j}$) between parenthesis.

4.6.2.1 LP worst-case workload of a task executing on c cores

Given a task τ_i, this step computes an array μ_i of size m, which includes the worst-case workload of τ_i when NPRs are distributed over c cores, being

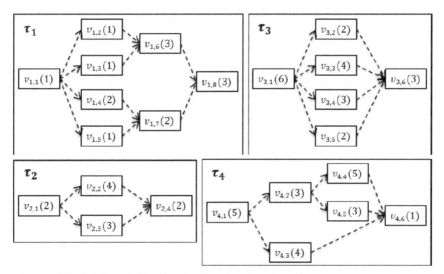

Figure 4.7 DAGs of lp(k) tasks; the $C_{i,j}$ of each node $v_{i,j}$ is presented in parenthesis.

c = {1,...,m} the index inside μ_i. Each element $\mu_i[c]$ is computed as follows:

$$\mu_i[c] = \sum {}^{parallel}_{c}\max\{C_{i,j}\}$$

where $max_c^{parallel}$ is the sum of the c largest NPRs of τ_i that can execute in parallel, maximizing the interference when using c cores. To this aim, the sum must consider the edges of τ_i's DAG to determine which NPRs can actually execute in parallel. Section 4.7.3 presents the algorithm that derives, for each NPR of τ_i, the set of NPRs from the same task that can potentially execute in parallel with it.

Table 4.1 shows the array μ_i for each of the tasks shown in Figure 4.7 with $m = 4$. For example, the worst-case workload $\mu_4 [2]$ occurs when NPRs $v_{4,3}$ and $v_{4,4}$ execute in parallel, with an overall impact of 9 time units. τ_2 has a maximum parallelism of 2, so $\mu_2 [3]$ and $\mu_2 [4]$ are equal to 0.

4.6.2.2 Overall LP worst-case workload

The lower-priority interference depends on how the execution of lp(k) is distributed across the m cores. We define $e^m = \{s_1,...,s_{p(m)}\}$ as the set of different execution scenarios (and so interference scenarios) of lp(k) running on m cores. p(m) is equal to the number of partitions[4] of m, and can be computed with the pentagonal number theorem from Euler's formulation:

$$\sum_q (-1)^q p\left(m - \frac{q(3q-1)}{2}\right)$$

where the sum is over all nonzero integers q (positive and negative) [22].

Table 4.1   Worst-case workloads of tasks in Figure 4.7

$\mu_1[c]$	$\mu_2[c]$	$\mu_3[c]$	$\mu_4[c]$
$C_{1,6}$ or $C_{1,8} = 3$	$C_{2,2} = 4$	$C_{3,1} = 6$	$C_{4,1}$ or $C_{4,4} = 5$
$C_{1,6} + C_{1,7} = 5$	$C_{2,2} + C_{2,3} = 7$	$C_{3,3} + C_{3,4} = 7$	$C_{4,4} + C_{4,3} = 9$
$C_{1,6} + C_{1,4} +$ $C_{1,5} = 6$	0	$C_{3,3} + C_{3,4} + C_{3,2}$ or $C_{3,5} = 9$	$C_{4,4} + C_{4,3} +$ $C_{4,5} = 12$
$C_{1,2} + C_{1,3} +$ $C_{1,4} + C_{1,5} = 5$	0	$C_{3,2} + C_{3,3} +$ $C_{3,4} + C_{3,5} = 11$	0

---

[4]In number theory and combinatory, a partition of a positive integer m is a way of writing m as a sum of positive integers. Two sums that differ only in the order of their summands are considered the same partition.

Table 4.2 the five possible execution scenarios assuming four cores [e_4, $p(4) = 5$]. The number of tasks being executed in each execution scenario s_l in e^m is given by its cardinality, i.e., $|s_l|$.

Each execution scenario s_l in e^m has an associated overall worst-case workload, computed as:

$$\rho_k[s_l] = \sum^{s_l} \max_{|s_l|}\{\mu_i\}$$

Where the right-hand side represents the sum of the $|s_l|$ largest combinations of μ_i that fits in the scenario s_l, and so maximizes the interference. Section 4.7.3 formulates the above equation as an ILP.

Table 4.3 shows the $\rho_k[s_l]$ of each execution scenario and the $\mu_i[c]$ considered in Table 4.1 and 4.2. For instance, the overall worst-case workload of s_3, $\rho_k[s3] = 19$ results when τ_4 executes on two cores ($\mu_4[2] = 9$), and τ_2 and τ_3 execute on one core each ($\mu_2[1] = 4$ and $\mu_3[1] = 6$).

4.6.2.3 Lower-priority interference

Finally, given the overall worst-case workload for each scenario $\mu_k[s_l]$, the lower-priority interference of lp(k) can be reformulated as the maximum overall worst-case workload among all scenarios:

$$\Delta_k^m = \max_{s_l \in e^m} \rho_k[s_l]$$

$$\Delta_k^{m-1} = \max_{s_l \in e^{m-1}} \rho_k[s_l]$$

Table 4.2 Five possible scenarios of taskset in Figure 4.7, assuming a four core system

$s_p \in e^4$	$|s_p|$	Execution scenario description
$s_1 = \{1,1,1,1\}$	4	Each task runs in 1 core
$s_2 = \{2,2\}$	2	Each task runs in 2 cores
$s_3 = \{2,1,1\}$	3	1 task runs in 2 cores and 2 task in 1 cores each
$s_4 = \{3,1\}$	2	1 task runs in 3 cores and 1 task in 1 core
$s_5 = \{4\}$	1	1 task runs in 4 cores

Table 4.3 Computed worst-case workload for each of the scenarios in Table 4.2

s_l	$\rho_k[s_l]$
s_1	$\mu_1[1] + \mu_2[1] + \mu_3[1] + \mu_4[1] = 18$
s_2	$\mu_2[2]$ or $\mu_3[2] + \mu_4[2] = 16$
s_3	$\mu_4[2] + \mu_2[1] + \mu_3[1] = 19$
s_4	$\mu_4[3] + \mu_3[1] = 18$
s_5	$\mu_3[4] = 11$

where the right-hand sides provide the maximum worst-case workload among e^m and e^{m-1} scenarios.

The lower-priority interference of lp(k) is given by the maximum $\rho_k[s_1]$, i.e., $\Delta_k^4 = 19$. On the contrary, the pessimistic approach selects the sum of the m largest NPRs among all lower-priority tasks, i.e., $\Delta_k^4 = C_{3,1} + C_{4,1} + C_{4,4} + C_{2,2} = 20$. The pessimism comes from the fact that nodes $v_{4,1}$ and $v_{4,4}$ cannot be executed in parallel. Similarly, $\Delta_k^3 = 15$, while the pessimistic approach gives $\Delta_k^3 = 16$.

Clearly, LP-ILP allows computing a tighter lower-priority interference, at the cost of increasing the complexity of deriving it, compared to the LP-max approach.

4.6.3 Computation of Response Time Factors of LP-ILP

We showed that the schedulability of a DAG-based task-set under LP-ILP can be checked in pseudo-polynomial time if, beside deadline and period, we can derive: (1) the worst-case workload generated by each lower-priority task τ_i (i.e., μ_i), and (2) the overall worst-case workload of lower-priority tasks for each execution scenario s_1 in e^m (i.e., $\rho_m[s_1]$). The former can be computed at compile-time for each task, and it is independent from the task-set; the latter requires the complete task-set knowledge, and is computed at system integration time. In this section, we present the algorithms to compute these factors.

4.6.3.1 Worst-case workload of τ_i executing on c cores: $\mu_i[c]$

$\mu_i[c]$ is determined by the set of c NPRs of τ_i that can potentially execute in parallel. As a first step, we identify for each NPR the set of potential parallel NPRs; then, we compute the interference of parallel execution when different numbers of cores are used.

(1) Computing the set of parallel NPRs: Given the DAG $G_i = (V_i, E_i)$, Algorithm 4.2 computes, for each NPR $v_{i,j}$ in V_i, the set of NPRs that can execute in parallel with it.

The algorithm takes as input the DAG of task τ_i, the topological order of G_i, and, for each node $v_{i,j}$, the sets:

1. SIBLING($v_{i,j}$), which contains the nodes which have a common predecessor with $v_{i,j}$;
2. SUCC($v_{i,j}$), which contains the nodes reachable from $v_{i,j}$; and

Algorithm 4.2 Parallel NPRs of τ_i

Input: *(1)* $G_i = (V_i, E_i)$; *(2)* TOPOLOGICAL-ORDER(G_i);
(3) SIBLING $(v_{i,j})$, SUCC $(v_{i,j})$, PRED$(v_{i,j})$ $\forall v_{i,j} \in V_i$
Output: *Par*$(v_{i,j})$, $\forall v_{i,j} \in V_i$
1: **procedure** PARALLEL-NPR
2:      **for each** $v_{i,j} \in V_i$ **do**
3:          $Par(v_{i,j}) \leftarrow \emptyset$
4:          **for each** $v_{i,l} \notin$ SIBLING $(v_{i,j})$ **do**
5:              **if** $(v_{i,j}, v_{i,l}) \notin E_i$ **and** $(v_{i,l}, v_{i,j}) \notin E_i$ **then**
6:                  $Succ \leftarrow$ SUCC $(v_{i,l}) \backslash$SUCC$(v_{i,j})$
7:                  $Par(v_{i,j}) \leftarrow Par(v_{i,j}) \cup \{v_{i,l}\} \cup Succ$
8:              **end if**
9:          **end for**
10:     **end for**
11:     **for each** $v_{i,j} \in$ TOPOLOGICAL-ORDER(G_i) **do**
12:         **for each** $v_{i,l} \in$ PRED$(v_{i,j})$ **do**
13:             $Pred \leftarrow Par(v_{i,l}) \backslash$ **PRED**$(v_{i,j})$
14:             $Par(v_{i,j}) \leftarrow Par(v_{i,j}) \cup Pred$
15:         **end for**
16:     **end for**
17: **end procedure**

---

3. PRED($v_{i,j}$), which contains the nodes from which $v_{i,j}$ can be reached. It outputs, for each $v_{i,j}$, the set Par($v_{i,j}$), containing the nodes that can execute in parallel with it.

The algorithm iterates twice over all nodes in V_i. The first loop (lines 2–10) adds to Par($v_{i,j}$) (line 7) the set of sibling nodes $v_{i,l}$ that are not connected to $v_{i,j}$ by an edge (line 5), and the nodes reachable from $v_{i,l}$ [SUCC($v_{i,l}$)], discarding those connected to $v_{i,j}$ by an edge (line 6). The second loop (lines 11–15), which traverses V_i in topological order, adds to Par($v_{i,j}$) (line 14) the set of nodes Par($v_{i,l}$) computed at line 7, being $v_{i,l}$ a node from which $v_{i,j}$ can be reached [$v_{i,l}$ in PRED($v_{i,j}$)]. From Par($v_{i,l}$) we discard the nodes from which $v_{i,j}$ can be reached (line 13).

As an example, consider node $v_{1,3}$ of τ_1 in Figure 4.7. The first loop iterates over the sibling nodes $v_{1,2}$, $v_{1,4}$, and $v_{1,5}$. None of them is connected to $v_{1,3}$ by an edge (lines 4 and 5); also, SUCC($v_{1,2}$) = $\{v_{1,6}, v_{1,8}\}$, SUCC($v_{1,4}$) = $\{v_{1,7}, v_{1,8}\}$, and SUCC($v_{1,5}$) = $\{v_{1,7}, v_{1,8}\}$. The algorithm discards from SUCC($v_{1,2}$) nodes $\{v_{1,6}, v_{1,8}\}$, since they are already included in SUCC($v_{1,3}$) (line 6). This is not the case of $v_{1,7}$ in SUCC($v_{1,4}$) and SUCC($v_{1,5}$). Hence, we obtain Par($v_{1,3}$) = $\{v_{1,2}, v_{1,4}, v_{1,5}, v_{1,7}\}$. The second loop does not add new nodes to $Par(v_{1,3})$ because the unique node from which $v_{1,3}$ can be reached is $v_{1,1}$, and $Par(v_{1,1})$ is empty. When the second loop examines node $v_{1,7}$, the two sets $Par(v_{1,4})$ and $Par(v_{1,5})$ are considered, since $v_{1,4},v_{1,5}$ in $PRED(v_{1,7})$. Then, nodes $v_{1,2}$, $v_{1,3}$, and $v_{1,6}$ are included in $Par(v_{1,7})$, since none of them belongs to $PRED(v_{1,7})$.

(2) Impact of parallel NPRs on c cores: For any task τ_i, we present an ILP formulation to compute $\mu_i[c]$, i.e., the sum of the c largest NPRs in V_i that, when executed in parallel, generate the worst-case workload.

Parameters: (1) c, i.e., the maximum number of cores used by τ_i; (2) $v_{i,j}$ in V_i; (3) q_{i+1}, i.e., the number of NPRs; (4) $C_{i,j}$; and (5) *IsPar*$_{i,j,k}$ in $\{0,1\}$, i.e., a binary variable that takes 1 if $v_{i,j}$ and $v_{i,k}$ can execute in parallel, 0 otherwise.

Problem variables: (1) b_j in $\{0,1\}$, i.e., a binary variable that takes the value 1 if $v_{i,j}$ is one of the selected parallel NPRs, 0 otherwise, and (2) $b_{j,k} = b_j$ OR b_k with $b_{j,k}$ in $\{0,1\}$; $j \neq k$, i.e., an auxiliary binary variable.

Constraints:

1. $\sum_{j=1}^{q_{i+1}} b_j = c$, i.e., only c NPRs can be selected;

2. $\sum_{j=1}^{q_{i+1}} \sum_{k=j+1}^{q_{k+1}} b_{j,k}$ *IsPar*$_{i,j,k} = c$, i.e., the selected NPRs can be executed in parallel; and

3. $b_{j,k} \geq b_j + b_k - 1$; $b_{j,k} \leq b_j$; $b_{j,k} \leq b_k$, i.e., auxiliary constraints used to model the logical *AND*.

Objective function: $\sum_{c=1}^{m} \sum_{\forall \tau_j \in lp(k)} w_i^c \mu_i^c$.

4.6.3.2 Overall LP worst-case workload of lp(k) per execution scenario s_l: $\rho_k[s_l]$

Given the set lp(k) and an execution scenario s_l in e^m, we present an ILP formulation to derive $\rho_k[s_l]$, that is, the overall worst-case workload generated by lp(k) under s_l.

Parameters: (1) lp(k); (2) m; (3) s_l; and (4) $\mu_i[c]$, for all τ_i in lp(k), for all $c = 1,\ldots,m$.

Problem variable: w_i^c, i.e., a binary variable that takes the value 1 on the selected $\mu_i[c]$ that contributes to the worst-case workload, 0 otherwise.

Constraints:

1. $\sum\limits_{c=1}^{m} \sum\limits_{\forall \tau_j \epsilon lp(k)} w_i^c = |s_l|$, i.e., the number of tasks contributing to the worst-case workload must be equal to the size of the execution scenario;

2. For all τ_i in lp(k), $\sum\limits_{c=1}^{m} w_i^c \leq 1$, i.e., each task can be considered at most in one scenario;

3. $\sum\limits_{\forall \tau_j \epsilon lp(k)} w_i^c \geq 1$, c in s_l, i.e., for each number of cores considered in s_l, there exist at least one $\mu_i[c]$ that is selected;

4. $\sum\limits_{c=1}^{m} \sum\limits_{\forall \tau_j \epsilon lp(k)} w_i^c c = m$, the number of cores considered is m.

Objective function: $max \sum\limits_{c=1}^{m} \sum\limits_{\forall \tau_j \epsilon lp(k)} w_i^c \mu_i^c.$

4.6.4 Complexity

The complexity of the response time analysis is still pseudo-polynomial. We hereafter discuss the complexity of the LP-ILP analysis.

Algorithm 4.2 requires specifying for each node in V_i the sets SIBLING, SUCC and PRED, which can be computed in quadratic time in the number of nodes. Similarly, the complexity of Algorithm 4.1 is quadratic in the size of the DAG task, i.e., $O(|V_k|^2)$. The ILP formulation to compute $\mu_i[c]$ is performed for each task (except for the highest-priority one), and the number of cores ranges from 2 to m, hence the complexity cost is O(nm) O(ilp_A). It is important to remark that Algorithm 4.2 (as well as its inputs) and the ILP that computes $\mu_i[c]$ are executed at compile-time for each task and are independent of the task-set and the system where they execute.

$\rho_k[s_l]$ is computed for the execution scenarios e^m and e^{m-1}, and for each task τ_k (except for the lowest-priority task τ_n), hence the complexity cost is: O(n p(m)) O(ilp_B) + O(n p(m−1)) O(ilp_B). The cost of solving both ILP formulations is pseudo-polynomial, if the number of constraints is fixed [23]. Our ILP formulations have fixed constraints, with a function cost of O(ilp_A) and O(ilp_B) depending on $|V_k|$ and (m n) respectively.

Therefore, the cost of computing $\rho_k[s_l]$ for e^m dominates the cost of other operations; hence, the complexity of computing the lower priority interference is pseudo-polynomial in the number of tasks and execution scenarios, i.e., cores.

4.7 Specializing Analysis for the Partitioned/Static Approach

The use of dynamic schedulers in certain high-criticality real-time systems may be problematic. In the automotive domain, for example, the static allocation of system components (named runnables in the AUTOSAR nomenclature) define a valid application configuration, for which the application is tested and validated. This configuration defines a specific data-flow, i.e., an order in which components process data, and an end-to-end latency between sensors and actuators, e.g., the gas pedal (sensor) and the injection (actuator). A dynamic allocation instead generates different data-flows and sensor-actuator latencies that may result in invalid configurations. The use of static allocation is therefore of paramount importance for these types of systems to guarantee the correct functionality.

In this section[5], a static allocation of parallel applications is proposed based on the OpenMP4 tasking model, in order to comply with the restrictive predictability requirements of safety-critical domains. An optimal task-to-thread mapping is derived based on an ILP formulation, providing the best possible response time for a given parallel task graph.

Two different formulations are proposed to optimally deal with both the tied and untied tasking models. Then, different heuristics are proposed for an efficient (although sub-optimal) task-to-thread mapping, with a reduced complexity. Experiments on randomly generated workloads and a real case-study are provided to characterize the worst-case response time of the proposed mapping strategies for each tasking model. The results show a significant reduction in the worst-case makespan with respect to existing dynamic mapping methods, taking a further step towards the adoption of OpenMP in real-time systems for an efficient exploitation of future embedded many-core systems.

4.7.1 ILP Formulation

This section proposes an Integer Linear Programming (ILP) formulation to solve the problem of optimally allocating OpenMP tasks to threads. The problem is to determine the minimum time interval needed to execute a given OpenMP application on m threads, both in the case of tied and untied tasks. In other words, we seek to derive the optimal mapping of task (or task parts) to threads so that the task-set makespan is minimized.

---

[5]This section was published as a conference paper at AspDAC [30].

The system model is the same as in the previous sections, with the following modifications needed to account for the OpenMP task semantics. An OpenMP application is modeled as an OpenMP-DAG G composed of N tasks $\tau_1, \ldots, \tau_N$. Each task τ_i is composed of n_i parts $P_{i,1}, \ldots, P_{i,ni}$. The Worst-Case Execution Time (WCET) of part $P_{i,j}$ of task τ_i is denoted as $C_{i,j}$. The total number of threads where tasks can be executed on a multi-core platform is denoted as m.

4.7.1.1 Tied tasks

The optimal allocation problem for tied tasks is modeled by starting from the set of tasks $\tau_1, \ldots, \tau_N$ and by adding a sink task τ_{N+1} with a single task part having null WCET (i.e., $C_{N+1,1} = 0$) and with incoming edges from the task parts without any successors in the original OpenMP-DAG.

The starting time of τ_{N+1} corresponds to the minimum completion time of the considered application; hence it represents our minimization objective.

Input parameters: (1) m: number of threads available for execution; (2) N: number of tasks in the system; (3) $C_{i,j}$: WCET of the j-th part of task τ_i; (4) G = (V, E): DAG representing the structure of the OpenMP application; (5) D: relative deadline of the OpenMP-DAG; (6) $succ_{i,j}$: set of immediate successors of part $P_{i,j}$ of τ_i; (7) rel_i: set of tasks having a relative relationship with τ_i (either as antecedents or descendants).

Problem variables: (1) $X_{i,k}$ in $\{0,1\}$: binary variable that is 1 if task τ_i is executed by thread k, 0 otherwise; (2) $Y_{i,j,k}$ in $\{0,1\}$: binary variable that is 1 if the j-th part of task τ_i is executed by thread k, 0 otherwise; (3) $\psi_{i,j}$: integer variable that represents the starting time of part $P_{i,j}$ of task τ_i (i.e., its initial offset in the optimal schedule); (4) $a_{i,j,w,z,k}$, $b_{i,w,k}$ in $\{0,1\}$: auxiliary binary variables.

Objective function: The objective function aims to minimize the starting time of the dummy sink task τ_{N+1}: min $\psi_{N+1,1}$ and represents the minimum makespan. A scheduling can be declared feasible if the minimum makespan is $\psi_{N+1,1} \leq D$.

Initial Assumptions: (i) The first part of the first task must begin at time $t = 0$: $\psi_{1,1} = 0$; (ii) The first task is executed by thread 1:

$$X_{1,1} = 1$$
$$X_{1,k} = 0 \quad \forall k \in \{2, \ldots, m\}$$
$$Y_{1,j,1} = 1 \quad \forall j \in \{1, \ldots, n_1\}$$
$$Y_{1,j,m} = 0 \quad \forall j \in \{1, \ldots, n_1\}, \forall k \in \{2, \ldots, m\}$$

Constraints

1. Each task is executed by only one thread:

$$\sum_{k=1}^{m} X_{i,k} = 1 \quad \forall_i \in \{1, \ldots, N\}$$

This constraint enforces the tied scheduling clause, i.e., for each task τ_i, only one binary variable $X_{i,k}$ is set to 1 among the m variables referring to the available threads.

2. All parts of each task are allocated to the same thread:

$$n_i \cdot X_{i,k} = \sum_{j=1}^{n_i} Y_{i,j,k} \quad \forall_i \in \{1, \ldots, N\}, \forall_k \in \{1, \ldots, m\}$$

This constraint establishes the correspondence between the $X_{i,k}$ and $Y_{i,j,k}$ variables.

3. All precedence requirements between task parts must be fulfilled:

$$\forall_i, \omega \in \{1, \ldots, N+1\}, \forall_j \in \{1, \ldots, n_i\},$$
$$\forall_z \in \{1, \ldots, n_w\} \,|\, P_{\omega,z} \in succ_{i,j},$$
$$\psi_{i,j} + C_{i,j} \leq \psi_{w,z}.$$

For each pair of task parts, if a precedence constraint connects them, then the latter cannot start until the former has completed execution. Notice that this constraint also applies to the sink task τ_{N+1}.

4. The execution of different task parts must be non-overlapping:

$$\forall_i, \omega \in \{1, \ldots, N\}, \forall_j \in \{1, \ldots, n_i\}, \forall_z \in \{1, \ldots, n_w\},$$
$$\forall_k \in \{1, \ldots, m\} \,|\, (\omega \neq i) \vee (j \neq z),$$
$$(Y_{i,j,k} = 1 \wedge Y_{w,z,k} = 1) \Rightarrow$$
$$(\psi_{i,j} + C_{i,j} \leq \psi_{w,z} \vee \psi_{w,z} + C_{w,z} \leq \psi_{i,j})$$

In other terms, if two task parts are allocated to the same thread, then either one finishes before the other begins, or vice versa. This constraint can be written as:

$$\forall i, \omega \in \{1, \ldots, N\}, \forall j \in \{1, \ldots, n_i\}, \forall z \in \{1, \ldots, n_w\},$$
$$\forall k \in \{1, \ldots, m\} \,|\, (\omega \neq i) \vee (j \neq z),$$
$$\psi_{i,j} + C_{i,j} \leq \psi_{w,z} + M(2 + a_{a,j,w,z,k} - Y_{i,j,k} - Y_{w,z,k})$$
$$\psi_{w,z} + C_{w,z} \leq \psi_{i,j} + M(3 - a_{a,j,w,z,k} - Y_{i,j,k} - Y_{w,z,k})$$

where M is an arbitrarily large constant. Indeed, if $a_{i,j,w,z,k} = 1$, then the first inequality is always inactive, while the second one is active only if $Y_{i,j,k} = 1$ and $Y_{w,z,k} = 1$. Similarly, if $a_{i,j,w,z,k} = 0$, then the first inequality is active only if $Y_{i,j,k} = 1$ and $Y_{w,z,k} = 1$, while the second one is always inactive.

5. The Task Scheduling Constraint 2 (TSC 2) as described in Chapter 3 must be satisfied:

$$\forall i, \omega \in \{1, \ldots, N\}, i \neq w, T_w \notin rel_i, \forall k \in \{1, \ldots, m\},$$
$$(X_{i,k} = 1 \wedge X_{w,z} = 1) \Rightarrow$$
$$(\psi_{i,n_i} + C_{i,n_i} \leq \psi_{w,1}) \vee (\psi_{w,n_w} + C_{w,n_w} \leq \psi_{i,1}).$$

This constraint imposes that one task cannot be allocated to a thread where another task that is neither a descendant nor an antecedent of the considered task is suspended. This is equivalent to saying that if two tasks not related by any descendance relationship are allocated to the same thread, then one of them must have finished before the other one begins. Therefore, the last task part of either task plus its WCET must be smaller than or equal to the starting time of the first task part of the other one. As for constraint (iv), it can be rewritten as:

$$\forall i, \omega \in \{1, \ldots, N\}, i \neq w, T_w \notin rel_i, \forall k \in \{1, \ldots, m\},$$
$$\psi_{i,n_i} + C_{i,n_i} \leq \psi_{w,1} + M(2 + b_{i,w,k} - X_{i,k} - X_{w,k})$$
$$\psi_{w,n_w} + C_{w,n_w} \leq \psi_{i,1} + M(3 - b_{i,w,k} - X_{i,k} - X_{w,k}).$$

Note that all constraints [except constraint (iii)] need not be applied to τ_{N+1}.

4.7.1.2 Untied tasks

The ILP formulation proposed for tied tasks can be applied for untied tasks with the following modifications. The initial assumption (ii) is replaced as follows: $Y_{1,1,1} = 1$.

Since different parts of the same task are allowed to be executed by different threads, constraints (i) and (ii) are replaced by:

$$\sum_{k=1}^{m} Y_{i,j,k} = 1 \forall i \in \{1, \ldots, N\}, \forall j \in \{1, \ldots, n_i\}$$

and the variables $X_{i,k}$ are no longer needed. Finally, constraint (v) does not apply for untied tasks and thus the auxiliary variables $b_{i,w,k}$ are not needed.

4.7.1.3 Complexity

The problem of determining the optimal allocation strategy of an OpenMP-DAG composed of untied tasks has a direct correspondence with the makespan minimization problem of a set of precedence-constrained jobs (task parts in our case) on identical processors (threads in a team in our case). This problem, also known as job-shop scheduling, has been proven to be strongly NP-hard by a result of Lenstra and Rinnooy Kan [18]. The complexity of the problem for the tied tasks cannot be smaller than in the untied case. Indeed, when each task has a single task part, the problem for tied tasks reduces to that for untied tasks.

In the presented ILP formulations for both the tied and untied tasks, the number of variables and the number of constraints grow as $O(N^2 p^2 m)$, where $p = \max_{i=1,\ldots,N} \quad n_i$.

Given the problem complexity and poor scalability of the ILP formulation, the next section proposes an efficient heuristic for providing sub-optimal solutions within a reasonable amount of time.

4.7.2 Heuristic Approaches

In the context of production scheduling, several heuristic strategies have been proposed to solve the makespan minimization problem of precedence constrained jobs on parallel machines [20, 24]. More specifically, different priority rules have been proposed in the literature to sort a collection of jobs subject to arbitrary precedence constraints on parallel machines. Such ordering rules allow selecting the next job to be executed in the set of ready jobs.

The ordering rules that have been shown to perform well in the context of parallel machine scheduling are [20, 24]:

1. **Longest Processing Time (LPT)**: The job with the longest WCET is selected;

2. **Shortest Processing Time (SPT)**: The job with the shortest WCET is selected;
3. **Largest Number of Successors in the Next Level (LNSNL)**: The job with the largest number of immediate successors is selected;
4. **Largest Number of Successors (LNS)**: The job with the largest number of successors overall is selected;
5. **Largest Remaining Workload (LRW)**: The job with the largest workload to be executed by its successors is selected.

We build upon such results to make them applicable to the considered problem. At any time instant, the set of ready jobs of a given instance of an OpenMP-DAG corresponds to the set of task parts that have not completed execution and whose precedence constraints are fulfilled.

This section presents an algorithm for allocating tied and untied task parts on the different threads following one of the above-mentioned ordering criteria, such that the partial ordering between task parts is respected.

4.7.2.1 Tied tasks

Algorithm 4.3 instantiates the procedure for the case of tied tasks, for which existing heuristic strategies cannot be directly applied. The algorithm takes the structure G of an OpenMP-DAG and the number of available threads m as inputs, and it outputs a heuristic allocation of tied OpenMP tasks to threads.

The idea behind the algorithm is to allocate ready task parts to the first available thread, following a pre-determined criterion to choose among ready tasks, while enforcing the specific semantics of the OpenMP tasking model. First, a list R of ready task parts is initialized with $P_{1,1}$, and an array L of size m with null initial values is used to store the last idle time on each thread (lines 2–3). The while loop at lines 4–25 iterates until all task parts have been allocated, i.e., until the size of list A, which contains the allocated jobs, reaches the total number of parts in the task-set. At each iteration, a new task part is allocated to one of the threads. Specifically, at line 5, the index k of the earliest available thread is determined by function FirstIdleThread. Then, the procedure NextReadyJob returns the ready task part $P_{i,j}$ selected according to one of the ordering rules described above. The allocation of the selected task part must always respect TSC 2. Hence, any time the first part of a new task is selected, the function must check its descendance relationships with the tasks currently suspended on thread k, stored in the list S_k. If $P_{i,j}$ is the first part of τ_i (line 7), then it is allocated on core k; otherwise, it is

Algorithm 4.3 Heuristic allocation of an OpenMP application comprising tied tasks

---

1: **procedure** HEURTIED(G, m)
2:     $A \leftarrow \emptyset; R \leftarrow P_{1,1}$
3:     $L \leftarrow \text{ARRAY}(m, 0) : S \leftarrow \text{ARRAY}(m, \emptyset)$
4:     **while** $\text{SIZE}(A)! = \sum_{i=1}^{N} n_i$ **do**
5:         $k \leftarrow \text{FIRSTIDLETHREAD}(L)$
6:         $P_{i,j} \leftarrow \text{NEXTREADYJOB}(k, R, S_k, G)$
7:         **if** $j == 1$ **then**
8:             $\theta_i \leftarrow k$
9:             **if** $j! = n_i$ **then**
10:                 $S_k \leftarrow \text{APPEND}(i, S_k)$
11:             **end if**
12:         **else if** $j == n_i$ **then**
13:             $S_k \leftarrow \text{REMOVE}(i, S_k)$
14:         **end if**
15:         $\psi_{i,j} = \max(L_{\theta_i}, \psi_{i,j}); L_{\theta_i,} \leftarrow L_{\theta_i} + C_{i,j}$
16:         $A \leftarrow \text{APPEND}(P_{i,j}, A); R \leftarrow \text{REMOVE}(P_{i,j}, R)$
17:         **for** $P_{k,z}|(P_{i,j}, P_{k,z}) \in E$ **do**
18:             **if** $\psi_{k,z} < \psi_{i,j} + C_{i,j}$ **then**
19:                 $\psi_{k,z} \leftarrow \psi_{i,j} + C_{i,j}; F_{k,z} = F_{k,z} + 1$
20:                 **if** $F_{k,z} == \text{SIZE}(\text{INEDGES}_{k,z})$ **then**
21:                     $R \leftarrow \text{APPEND}(P_{k,z}, R)$
22:                 **end if**
23:             **end if**
24:         **end for**
25:     **end while**
26:     **return** $\max_{i=1}^{m} L_i$
27: **end procedure**

---

allocated on thread θ_i, according to the tied scheduling clause. Also, if that task part is not the final one (line 9), τ_i is appended to the list of tasks currently suspended on thread k. Otherwise, if $P_{i,j}$ is the final part of τ_i (line 12), τ_i can be removed from the list of tasks currently suspended on thread k. In both cases, the starting time of $P_{i,j}$ is updated, as well as the last idle time on thread k (line 15). In addition, $P_{i,j}$ is added to the list of allocated jobs and removed from the list of ready jobs (line 16). Once $P_{i,j}$ has been allocated, other jobs may become ready. All the successors of $P_{i,j}$ are scanned and an internal counter ($F_{k,z}$) is incremented for each vertex (for loop at lines 17–24). Once the counter reaches the number of its immediate predecessors, the task part may be appended to the list of ready vertices (line 21). Finally, the makespan corresponding to the generated allocation is returned. At the end of the algorithm, $\psi_{i,j}$ stores the starting time of any part $P_{i,j}$ in the final schedule, and θ stores the mapping of tasks to threads.

The algorithm runs in polynomial time in the size of the task-set; specifically, the time complexity is $O\left(\left(\sum_{i=1}^{N} n_i\right)^2\right)$.

4.7.2.2 Untied tasks

Algorithm 4.3 can be applied also in the case of untied tasks with some simplifications. In particular, the function NextReadyJob does not need to check the validity of TSC 2. Hence, the array S is not required, and all the operations on S at lines 7–14 do not need to be performed. On the other hand, the algorithm must keep track of the thread associated to each task part (instead of each task).

4.7.3 Integrating Interference from Additional RT Tasks

We now generalize the static setting by considering a set of n OpenMP applications modeled as a collection of OpenMP DAGs $\Gamma = \{G_1,\dots,G_n\}$. Each DAG is released sporadically (or periodically) and has a relative deadline D_i, which is constrained to be smaller than or equal to its corresponding period (or inter-arrival time) T_i.

We assume that parts of each tasks are statically partitioned to the m available threads. At any time instant, the scheduler selects among the ready task parts the one that should be executed by a given thread according to partitioned fixed-priority preemptive scheduling. In addition, we assume that OpenMP applications are statically prioritized, i.e., each DAG G_i is associated with a unique (fixed) priority that is used by the scheduler to select which task parts should be executed at any time instant by any of the threads.

In order to compute an upper-bound on the response time R_i of a given OpenMP-DAG G_i, we proceed by computing an upper-bound on the response time of each task part in the OpenMP-DAG, following a predefined order dictated by any topological sorting of the DAG. At each step, the response time of the considered vertex is computed considering all its immediate predecessors, one at a time. A safe upper-bound on the response time of the vertex under analysis will be selected as the maximum of such values. The maximum response time among vertices without successors will be selected as upper-bound to the response time of the DAG-task G_i.

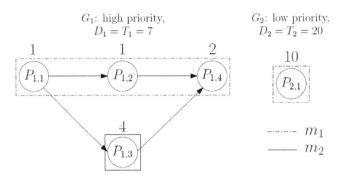

Figure 4.8   Tasks example.

4.7.4 Critical Instant

We hereafter prove that the synchronous periodic arrival pattern does not represent the worst-case release sequence for the OpenMP-DAG task model assumed. Consider a task-set composed of two OpenMP-DAG tasks G_1 and G_2, whose structure and parameters are illustrated in Figure 4.8. The figure also reports the static allocation of task parts to threads: parts $P_{1,1}$, $P_{1,2}$, $P_{1,4}$, and $P_{2,1}$ are allocated to thread m_1, while part $P_{1,3}$ is allocated to thread m_2.

We can immediately see that $R_1 = 7$, as G_1 is the highest priority RT task in the system. In order to compute the response time of G_2, we focus on thread m_1 and first consider the synchronous periodic arrival pattern for G_1, which produces the schedule in Figure 4.9a and yields a response time of 21 time units for G_2. However, if we consider the release pattern in Figure 4.9b, where the release of G_1 has an offset of two time units, we observe that the response time of G_2 becomes equal to 23.

This example shows that it is very difficult to exactly quantify the interference a task may suffer from higher-priority tasks in the worst-case. This is mainly due to the precedence constraints between parts of the same tasks, and to the fact that any vertex is allowed to execute on its corresponding thread only when all its predecessors (possibly allocated to different threads) have completed their execution. In order to overcome these problems, we derive a safe upper-bound on the response time of a given task by considering the densest possible packing of jobs generated by a legal schedule in any time interval. Specifically, we consider a pessimistic scenario (see Figure 4.9c):

- the first instance of a higher-priority task is released as late as possible;
- subsequent instances are released as soon as possible;
- higher-priority jobs are considered as if precedence constraints were removed (their WCET is "compacted").

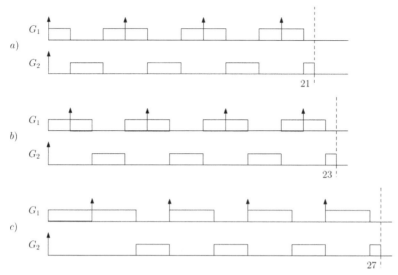

Figure 4.9 Different release patterns for the example of Figure 4.8. (a) represents the most optimistic case, while (c) the most pessimistic, i.e., yelding to the highest WCET. (b) represents an intermediate case.

4.7.5 Response-time Upper Bound

Algorithm 4.4 computes an upper-bound on the response time of an OpenMP-DAG by considering the above-described pessimistic scenario leading to the densest possible packing of higher-priority task parts:

The function SELFINTERFERENCE calculates the self-interference suffered by task part $P_{k,i}$ as the sum of the WCETs of all parts $P_{k,j}$ belonging to the same task and such that:

1. they are allocated to the same thread as $P_{k,i}$;
2. there is no path starting at $P_{k,i}$ that can reach $P_{k,j}$;
3. there is no path starting at $P_{k,j}$ that can reach $P_{k,i}$.

With the above algorithm in place, different heuristics can be proposed to find a feasible allocation of task parts to threads/cores. Among the ones we tried, we found that the best schedulability performances are obtained with a Best Fit approach that works as follows:

- It assigns RT tasks in non-increasing priority order, i.e., starting from the highest priority task and moving towards lower priority ones.
- For each task it defines a topological order for all task parts.

Algorithm 4.4 Upper-bound on the response time of an OpenMP-DAG by considering the densest possible packing of higher-priority task parts

```
1: procedure DENSESTPACKINGALG($G_k$)
2:     $\sigma \to$ TOPOLOGICALORDER($G_k$)
3:     for $P_{k,i} \in \sigma$ from source to sink do
4:         $R_{max} = \max_{j \in \mathrm{PRED}(k,i)} R_{k,j}$
5:         S = SELFINTERFERENCE($k,i$)
6:         $R \leftarrow C_{k,i} + S$
7:         $R_{prev} \leftarrow 0$
8:         while $R \neq R_{prev}$ do
9:             $R_{prev} \leftarrow R$
10:            $R \leftarrow C_{k,i} + S$
11:            for $P_{h,j}$ such that $h < k$ and $\theta_{h,j} == \theta_{k,i}$ do
12:                $R \leftarrow R + \left\lceil \frac{R_{prev} + R_{h,j} - C_{h,j}}{T_h} \right\rceil C_{h,j}$
13:            end for
14:        end while
15:        If $R_{max} + R > D_k$ then
16:            $sched \leftarrow 0$
17:            break
18:        else
19:            $sched \leftarrow 1$
20:            $R_{k,i} \leftarrow R_{max} + R$
21:        end if
22:    end for
23:    return $\{sched, R_{k,sink}\}$
24: end procedure
```

- Following the topological order, each task part is assigned to the core that minimizes its partial response time, i.e., the response time of the RT task until the considered task part.
- If any of the considered task parts has a partial response time that exceeds its relative deadline, the algorithm fails, declaring the RT task-set not schedulable.

The partial response time of each task part can be easily computed using Algorithm 4.4, executing the operations within the for loop at line 3. Once the selection is made for a task part, there is no need to recheck the schedulability of the parts already assigned belonging to higher priority tasks, since this last assignment does not interfere with them. However, it is necessary to reconsider the task parts belonging to the same RT task that may experience an increase in the interference. The only task parts that may be affected by the last task part assigned are those that have no precedence constraints with it. For these ones, we re-compute their partial response-time after the new assignment. Since there is no backtracking in this case, the complexity of the heuristic remains reasonable, at the penalty of some added pessimism.

4.8 Scheduling for I/O Cores

This paragraph briefly describes the scheduler adopted at host level, i.e., in the I/O cores.

According to system requirements, the OS running on the host processor must be Linux. Moreover, the Linux kernel must be patched with the PREEMPT_RT patch[6]. This is an on-going project supported by the OSADL association[7] to add real-time performance to the Linux kernel by reducing the *maximum* latency experienced by an application, mainly through preemptible spinlocks and in-thread interrupt management (See also essential work in [25–38]). The patch makes the system more predictable and deterministic; however, it often increases the *average* latency. Currently, the patch only partially works on the reference platform due to missing support for SMP in the Linux kernel; full support will be added during the next months. Concerning the scheduling policy, the OS must provide a fixed-priority preemptive FIFO-scheduling algorithm. Therefore, the basic scheduling algorithm will be the SCHED_FIFO policy specified by the POSIX standard. The optional requirement R5.21 suggests to have a Linux kernel higher than 3.14 for investigating potential benefits given by the dynamic-priority SCHED_DEADLINE Linux scheduler. This possibility will be explored at a later stage of the project. Access to shared resources in the host cores is handled through the Priority Inheritance (PI) policy provided by the Linux kernel.

4.9 Summary

In this chapter, we described the design choices related to the implementation of a partitioned scheduler for allocating the computing resources to the different threads in the system. In particular, we detailed the thread model adopted in the project, and the local scheduler adopted at core level, based on fixed thread priorities.

---

[6]PREEMPT_RT Linux patch, https://rt.wiki.kernel.org
[7]OSADL, Open Source Automation Development Lab, http://www.osadl.org/

Such a scheduler has then been enhanced with the enforcement of a limited pre-emption scheduling policy that corresponds to the execution model supported by the OpenMP tasking model, as well as allowing increasing the predictability of the analysis, without sacrificing the schedulability. According to the limited pre-emption scheduling model, each thread can be pre-empted only at particular pre-emption points. The framework provides a method to compute the length of the largest non-preemptive region that can be tolerated by each thread (at each different priority). Then, threads execute along non preemptive regions. In a generic model such as the one introduced in this chapter, this means inserting the minimum possible number of preemption points such that the schedulability of higher priority thread is not affected. Of course, specifying this model so that it adheres to OpenMP semantics means that the identification of these preemption points exploits information inherited from the OpenMP task semantics, i.e., OpenMP TSPs will be used as potential candidates.

We then described the implementation of an enhanced global scheduler with migration support. Such a scheduler is integrated with the OpenMP dynamic mapping policy to allow for a work-conserving resource allocation of computing resources. The scheduler adopts a cluster-wide ready queue where threads are ordered according to their priorities. Preemptions are allowed only at task-part boundaries when a TSP is reached. TSPs are also natural polling points to deal with new incoming offloads without requiring interrupts.

The task model adopted, namely the cp-task model, generalizes the classic sporadic DAG task model by integrating conditional branches. The topological structure of a cp-task graph has been formally characterized by specifying which connections are allowed between conditional and non-conditional nodes. Then, a schedulability analysis has been derived to compute a safe upper-bound on the response-time of each task in pseudo-polynomial time. Besides its reduced complexity, the proposed analysis has the advantage of requiring only two parameters to characterize the complex structure of the conditional graph of each task: the worst-case workload and the length of the longest path. Algorithms have also been proposed to derive these parameters from the DAG structure in polynomial time. Simulation experiments carried out with randomly generated cp-task workloads and real test-cases clearly showed that the proposed approach is able to improve over previously proposed solutions for tightening the schedulability analysis of sporadic DAG task systems. The first formulation of the analysis considered a full-preemption model (see [18]). Then, it has been extended to limited preemptive scheduling [24], and, finally, it has been specialized also for non-conditional DAGs [20, 29].

In this chapter, two methods have been proposed to compute the lower-priority interference: (1) a pessimistic but easy-to-compute method, named LP-max, which upper bounds the interference by selecting the NPRs with the longest worst-case execution time; and (2) a tighter but computationally-intensive method, named LP-ILP, which also takes into account precedence constraints among DAGs nodes in the analysis. Our results demonstrate that LP-ILP increases the accuracy of the schedulability test with respect to LP-max when considering DAG-based task-sets with different levels of parallelism.

The chapter then proposed an ILP formulation to derive an optimal static allocation compliant with the OpenMP4 tied and untied tasking model. With the objective of reducing the complexity of the ILP solver, five heuristics have been proposed for an efficient (although sub-optimal) allocation. Results obtained on both randomly generated task-sets and the 3DPP application (from the avionics domain) show a significant reduction in the worst-case makespan with respect to an existing schedulability upper-bound for untied tasks. Moreover, the proposed heuristics perform very well, closely matching the optimal solutions for small task-set, and outperforming the best feasible solution found by our ILP (after running the solver for a certain amount of time) for large task-sets and the 3DPP.

References

[1] Liu, C., Layland, J., Scheduling algorithms for multiprogramming in a hard real-time environment. *J. ACM* 20, 46–61, 1973.

[2] Leung, J. Y. T., Whitehead, J., On the complexity of fixed-priority scheduling of periodic, real-time tasks. *Perform. Eval.* 2, 237–250, 1982.

[3] Buttazzo, G., Bertogna, M., Yao, G., "Limited preemptive scheduling for real-time systems: a survey." *IEEE Transactions on Industrial Informatics*, 9, 3–15, 2013.

[4] Lehoczky, J., Sha, L., Ding, Y., "The rate monotonic scheduling algorithm: Exact characterization and average case behavior." In *Proceedings of the Real-Time Systems Symposium—1989, pp. 166–171. IEEE Computer Society Press*, Santa Monica, California, USA, 1989.

[5] Dhall, S. K., and Liu, C. L. On a real-time scheduling problem. *Operat. Res.*. 26, 127–140, 1978.

[6] Baruah, S., Cohen, N., Plaxton, G., and Varvel, D., Proportionate progress: A notion of fairness in resource allocation. *Algorithmica* 15, 600–625, 1996.

[7] Anderson, A., and Srinivasan, A., "Pfair scheduling: Beyond periodic task systems." In *Proceedings of the International Conference on Real-Time Computing Systems and Applications (Cheju Island, South Korea)*, IEEE Computer Society Press, 2000.

[8] Zhu, D., Mosse, D., and Melhem, R. G., "Multiple-resource periodic scheduling problem: how much fairness is necessary?" *24th IEEE Real-Time Systems Symposium (RTSS)* (Cancun, Mexico), 2003.

[9] Cho, H., Ravindran, B., and Jensen, E. D., "An optimal real-time scheduling algorithm for multiprocessors," *27th IEEE Real-Time Systems Symposium (RTSS)* (Rio de Janeiro, Brazil), 2006.

[10] Andersson, B., and Tovar, E., "Multiprocessor scheduling with few pre-emptions." In *Proceedings of the International Conference on Real-Time Computing Systems and Applications. (RTCSA)*, 2006.

[11] Funaoka, K., Kato, S., and Yamasaki, N., "Work-conserving optimal real-time scheduling on multiprocessors." In *Proceedings of the Euromicro Conference on Real-Time Systems*, 13–22, 2008.

[12] Funk, S., and Nadadur, V., "LRE-TL: An optimal multiprocessor algorithm for sporadic task sets." In *Proceedings of the Real-Time Networks and Systems Conference*, 159–168, 2009.

[13] Levin, G, Funk, S., Sadowski, C., Pye, I., Brandt, S, "DP-Fair: A Simple Model for Understanding Multiprocessor Scheduling." In *Proceedings of the 22nd Euromicro Conference on Real-Time Systems (ECRTS)*, Brussels, Belgium, pp. 1–10, 2010.

[14] Funk, S., Nelis, V., Goossens, J., Milojevic, D., Nelissen, G., and Nadadur, V., On the design of an optimal multiprocessor real-time scheduling algorithm under practical considerations (extended version). arXiv preprint arXiv:1001.4115, 2010.

[15] Nelissen, G., Su, H., Guo, Y., Zhu, D., Nelis, V., and Goossens, J., An optimal boundary fair scheduling. *Real-Time Sys. J.* 2014.

[16] Regnier, P., Lima, G., Massa, E., Levin, G., and Brandt, S., "RUN: Optimal multiprocessor real-time scheduling via reduction to uniprocessor," *IEEE 32nd Real-Time Systems Symposium (RTSS)*, 2011.

[17] Fisher, N., Goossens, J., and Baruah, S., Optimal online multiprocessor scheduling of sporadic real-time tasks is impossible. *Real-Time Sys. J.* 45.1-2, 26–71, 2010.

[18] Melani, A., Bertogna, M., Bonifaci, V., Marchetti-Spaccamela, A., and Buttazzo, G., "Response-Time Analysis of Conditional DAG Tasks in Multiprocessor Systems," in *27th Euromicro Conference on Real-Time Systems, ECRTS 2015*, Lund, Sweden, pp. 7–10, 2015.

[19] P-SOCRATES Deliverable 3.3.2. *Enhanced scheduler with migration support*. Delivery date: 31 March 2016.

[20] Serrano, M. A., Melani, A., Bertogna, M., and Quiñones, E., "Response-Time Analysis of DAG Tasks under Fixed Priority Scheduling with Limited Preemptions," in the *Design, Automation, and Test in Europe conference (DATE)*, Dresden, Germany, pp. 14–18, 2016.

[21] P-SOCRATES Deliverable 4.2.2. *Interference Model*. Delivery date: 31 March 2016.

[22] P-SOCRATES Deliverable 1.5.2. *Integrated Tool-chain*. Delivery date: 31 March 2016.

[23] Blumofe, R. D., and Leiserson, C. E., Scheduling multithreaded computations by work stealing. *J. ACM* 46, 720–748, 1999.

[24] Serrano, M. A., Melani, A., Kehr, S., Bertogna, M., and Quiñones, E., "An Analysis of Lazy and Eager Limited Preemption Approaches under DAG-based Global Fixed Priority Scheduling," in the *19th IEEE International Symposium on Object/Component/Service-oriented Real-time Distributed Computing (ISORC)*, Toronto, Canada, pp. 16–18, 2017.

[25] Anderson, T. E., "The performance of spin lock alternatives for shared-memory multiprocessors." In *IEEE Transactions on Parallel and Distributed Systems*, 1990.

[26] Craig, T. S., "Queuing spin lock algorithms to support timing predictability." In *Proc. Real-Time Sys. Symp.* pp. 148–157, 1993.

[27] Shen, C., Molesky, L. D., and Zlokapa, G., "Predictable synchronization mechanisms for real-time systems." In *Real-Time Systems*, 1990.

[28] Graunke, G., and Thakkar, S., "Syncronization algorithms for shared-memory multiprocessors." In *IEEE Computer*, 1990.

[29] Melani, A., Serrano, M. A., Bertogna, M., Cerutti, I., Quiñones, E., and Buttazzo, G., "A static scheduling approach to enable safety-critical OpenMP applications," in the *21st Asia and South Pacific Design Automation Conference (ASP-DAC)*, Tokyo, Japan, pp. 16–19, 2017.

[30] P-SOCRATES Deliverable 3.1. *Resource Allocation Requirements*. Delivery date: 30 April 2014.

[31] P-SOCRATES Deliverable 5.2. *Operating Systems Support Prototypes*. Delivery date: 31 March 2015.

[32] P-SOCRATES Annex I – *Description of Work*, 2014.

[33] P-SOCRATES Deliverable 3.2. *Mapping Strategies*. Delivery date: 31 March 2015.

[34] Joseph, M., Pandya, P., Finding response times in a real-time system. *Comput. J.* 29, 390–395, 1986.

[35] Maia, C., Nogueira, L., and Pinho, L. M., "Scheduling Parallel Real-Time Tasks using a Fixed-Priority Work-Stealing Algorithm on Multi-processors," in 8th *IEEE Symposium on Industrial Embedded Systems*, Porto, Portugal, pp. 19–21, 2013.

5

Timing Analysis Methodology

Vincent Nélis, Patrick Meumeu Yomsi and Luís Miguel Pinho

CISTER Research Centre, Polytechnic Institute of Porto, Portugal

This chapter focuses on the analysis of the timing behavior of software applications that expose real-time (RT) requirements. The state-of-the-art methodologies to timing analysis of software programs are generally split into four categories, referred to as *static*, *measurement-based*, *hybrid*, and *probabilistic analysis* techniques. First, we present an overview of each of these methodologies and discuss their advantages and disadvantages. Next, we explain the choices made by our proposed methodology in Section 5.2 and present the details of the solution in Section 5.3. Finally, we conclude the chapter in Section 5.4 with a summary.

5.1 Introduction

Most of the timing analysis tools focus only on determining an upper-bound on the Worst-Case Execution Time (WCET) of a program or function code that runs in isolation and without interruption. In other words, these tools do not consider all the interferences that the execution of the analyzed code may suffer when it runs concurrently with other tasks or programs on the same hardware platform. They typically ignore all execution interferences due to the contention for shared software resources (e.g., data shared between several tasks) and shared hardware resources (e.g., shared interconnection network)[1] [1]. Interferences from the operating system (OS) which frequently re-schedules and interrupts the programs are also ignored by WCET analyzers. All these interactions between the analyzed task, the OS, and all the

---

[1]Note that the OTAWA timing analysis tool is able to analyze parallel code with synchronization primitives [1].

other tasks running in the system are assessed separately and sometimes they are incorporated into a higher-level schedulability analysis. For the timing requirements to be fulfilled, it is neither acceptable nor realistic to ignore these sources of contention and interference at the schedulability-analysis level.

WCET analysis can be performed in a number of ways using different tools, but the main methodologies employed can be classified into four categories:

1. Static analysis techniques
2. Measurement-based analysis techniques
3. Hybrid analysis techniques
4. Measurement-based probabilistic analysis techniques

Note that the first three methodologies are usually acknowledged as equally important and efficient as they target different types of applications. In addition, they are not comparable in the sense that one technique has not been proven to dominate the others. The fourth technique is more recent and thus fewer results are available.

Measurement-based techniques are suitable for software that is less time-critical and for which the average-case behavior (or a rough WCET estimate) is more meaningful or relevant than an accurate estimate like, for example, in systems where the worst-case scenario is extremely unlikely to occur. For highly time-critical software, where every possible execution scenario must be covered and analyzed, the WCET estimate must be as reliable as possible and *static* or *hybrid* methods are therefore more appropriate. *Measurement-based probabilistic analysis* techniques are also designed for safety-critical systems to derive safe estimated execution time bounds, but they are not yet sufficiently mature to report on their efficiency and applicability. Indeed, a consensus is still to be reached in the research community on this matter.

For the execution time of a single *sequential* program run *in isolation*, Figure 5.1 shows how different timing estimates relate to the WCET and best-case execution time (BCET). The example program has a variable execution time that depends on (1) its input parameters and (2) its interactions with the system resources. The darker curve shows the actual probability distribution of its execution time; its minimum and maximum are the BCET and WCET respectively. The lower grey curve shows the set of execution times that have been observed and measured during simulations, which is a subset of all executions; its minimum and maximum are the *minimal measured time* and *maximal measured time*, respectively. For both static analysis tools

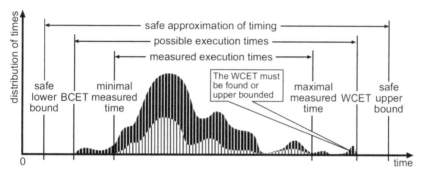

Figure 5.1 Example distribution of execution time (picture taken from [2]).

and measurements-based tools, in most cases the program state space and the hardware complexity are too large to exhaustively explore all possible execution scenarios of the program. This means that *the measured times are likely to be optimistic* and the *estimated times are likely to be pessimistic* – i.e., the measured times will in many cases overestimate the actual BCET and underestimate the actual WCET, while the approximated estimated times will in many cases underestimate the actual BCET and overestimate the actual WCET.

The next four subsections introduce each of the four timing-analysis methodologies and discuss their potential advantages and disadvantages.

5.1.1 Static WCET Analysis Techniques

Static WCET analysis is usually performed in three conceptual and possibly overlapping phases.

1. A flow analysis phase in which information about the possible program execution paths is derived. This step builds a control flow-graph from the given program with the aim of identifying the worst path (in terms of execution time).
2. A low-level analysis phase during which information about the execution time of atomic parts of the code (e.g., instructions, basic blocks, or larger code sections) is obtained from a model of the target architecture.
3. A final calculation phase in which the derived flow and timing information are combined into a resulting WCET estimate.

Flow analysis mostly focuses on loop bound analyses, hence upper-bounds on the number of iterations in each looping structure must be known to derive WCET estimates. Similarly, recursion depth must also be bounded.

Automatic methods to find these bounds have been proposed by the research community but for many available tools, some annotations on the maximum number of iterations in a loop must be provided manually in the code of the tasks by the application developer. Another purpose of flow analysis is to identify infeasible execution paths, which are paths that are executable according to the control-flow graph but are not feasible when considering the semantics of the program and the possible input data values. Discarding unfeasible paths at an early stage of the analysis considerably reduces the search space when trying to identify the longest path.

Low-level analysis methods typically use models of all the hardware components and their arbitration policies, including CPU caches, cache replacement policies, write policies, instruction pipeline, memory bus and their arbitration policies, etc. These models are typically expressed in the form of complex mathematical abstractions for which a worst-case operation can be estimated.

Pros: There are a few advantages of using static analysis techniques that rely on mathematical models.

- It eliminates the need for having the actual hardware available, which removes the cost of acquiring and setting up the target platform.
- It enables safe WCET upper-bounds to be derived without running the program on the target platform while still considering the influence of the state changes in the underlying hardware [3]. State changes include, e.g., a cache line being evicted, a pipeline being totally flushed out, etc.

Cons: On the downside, we shall note the following drawbacks.

- These approaches rely heavily on having an accurate model of the timing behavior of all the target hardware components and management policies, including modeling features like pipelines and caches that substantially affect the execution time of the task being executed. Although the embedded market used to be traditionally dominated by simple and predictable processors (which used to be moderately "easy" to model and allowed for deriving safe and tight bounds), with the increased computational needs of modern embedded systems, designers have moved to more complex processors which are now mainly designed for performance and not for predictability. For this new generation of processors, designing an accurate hardware model is very challenging, as all the intricacies contributing to the variation in the task execution times (e.g., caches, pipelines, out-of-order execution, branch prediction, automatic hardware prefetching, etc.) should be captured by the model to provide safe and sufficiently tight bounds. Because it is hardly feasible to accurately model all these acceleration mechanisms and their operation, static methods typically forbid their use and are struggling to adapt to modern hardware architectures.

- Besides the difficulty of modeling all these performance-enhancement hardware features, it must also be noted that generally, chip manufacturers do not publish the details of their internal workings, which further complicates/makes impossible the design of an accurate model.

- Although static approaches have the advantage of providing safe WCET bounds, they can be very pessimistic at times. This is because generally, each hardware resource is modeled separately, and all the worst-case estimates are then composed together to form the final WCET bound. However, at runtime, it is often impossible for all these individual worst-case scenarios to happen at the same time.

- The hardware model must be thoroughly verified to ensure that it indeed reflects the target hardware; failing to capture inherent performance enhancing features may result in overestimations of the execution times, whereas capturing all system states in a complex machine may lead to unacceptably long analysis times. Building and verifying the timing model for each processor variant is expensive, time consuming, and error prone. Custom variants and different versions of processors often have subtly different timing behaviors, rendering timing models either incorrect or unavailable.

It is very important to stress at this point that static analysis techniques have been designed primarily to analyze simple software codes meant to run on simple and predictable hardware architectures. These targeted codes are typically implemented by using high-level programming languages and by obeying strict and specific coding rules to reduce the likelihood of programmer error.

The modeling framework adopted by static analysis lends itself to formal proofs which help in establishing whether the obtained results are safe. Today, there are several static WCET tools that are commercially available, including *aiT* [4] and *Bound-T* [5]. Note that Bound-T is no longer actively developed due to both commercial and technical reasons. We redirect the interested reader to their website (http://www.bound-t.com/) for further details on this matter. There also exist several research prototypes, including *Chronos* [6], developed at National University of Singapore, *Heptane* [7], developed at the French National Institute for Research in Computer Science and Control (INRIA) IRISA in France, *SWEET* [8], developed at Mälardalen Real-Time Research Center (MRTC) in Sweden, and OTAWA [9] from IRIT in France.

5.1.2 Measurement-based WCET Analysis Techniques

The traditional and most common method in the industry to determine program timing is by measurements. The basic principle of this method follows the mantra that "the processor is the best hardware model." The program is executed many times on the actual hardware, with different inputs and *in isolation*, and the execution time is measured for each run by instrumenting the source code at different points [10]. Each measurement run exercises only one execution path throughout the program, and thus for the same set of input values, several thousands of program runs must be carried out to capture variations in execution time due to the fluctuation in system states. For those measurement-based approaches, the main challenge is essentially to identify the set of input arguments of the application that leads to its WCET.

Pros:

- Measurements are often immediately at the disposal of the programmer, and are useful mainly when the average case-timing behavior or an approximate WCET value is of interest.
- Most types of measurements have the advantage of being performed on the actual hardware, which avoids the need to construct a hardware model and hence reduces the overall cost of deriving the estimates.

Cons:

- Measurements require that hardware is available, which might not be the case for systems for which the hardware is developed in parallel with the software.
- It may be problematic to set up an environment which acts like the final system.
- The integrity of the actual code to be deployed in the target hardware is somehow depleted by the addition of the intrusive instrumentation code to measure the time, i.e., the measurements themselves add to the execution time of the analyzed program. This problem can be reduced, e.g., by using hardware measurement tools with no or very small intrusiveness, or by simply letting the added measurement code (and thus the extra execution time) remain in the final program. When doing measurements,

possible disturbances, e.g., interrupts, also have to be identified and compensated for.

- For most programs, the number of possible execution paths is too large to do exhaustive testing and therefore, measurements are carried out only for a subset of the possible input values, e.g., by giving potential "nasty" inputs which are likely to provoke the WCET, based on some manual inspection of the code. Unfortunately, the measured times will in many cases underestimate the WCET, especially when complex software and/or hardware are being analyzed. To compensate for this, it is common to add a safety margin to the worst-case measured timing, in the hope that the actual WCET lies below the resulting WCET estimate. The main issue is whether the extra safety margin provably provides a safe bound, since it is based on some informed estimates. A very high margin will result in resource over-dimensioning, leading to very low utilization while a small margin could lead to an unsafe system.

5.1.3 Hybrid WCET Techniques

Hybrid approaches, as the name implies, present the advantages of both static and measurement-based analysis techniques. Firstly, they borrow the flow-analysis phase from static methods to construct a control flow-graph of the given program and identify a set of feasible and potentially worst execution paths (in terms of execution time). Next, unlike static methods that use mathematical models of the hardware components, hybrid tools borrow their second phase from measurement-based techniques and determine the execution time of those paths by executing the application on the target hardware platform or by cycle-accurate simulators. To do so, the source code of the application is instrumented with expressions (instrumentation points) that indicate that a specific section of code has been executed. These instrumentation points are typically placed along the paths identified in the first phase as leading to a WCET. The application is then executed on the target hardware platform or on the simulator to collect execution traces. These traces are a sequence of time-stamped values that show which parts of the application has been executed. Finally, hybrid tools produce performance metrics for each part of the executed code and, by using the performance data and knowledge of the code structure, they estimate the WCET of the program.

Pros:

- Hybrid approaches do not rely on complex abstract models of the hardware architecture.

- They generally provide safe WCET estimates (i.e., higher than the actual WCET) and those are very often tighter than the estimates returned by static approaches (i.e., closer to the actual WCET).

Cons:

- The uncertainty of covering the worst-case behavior by the measurement remains since it cannot be guaranteed that the maximum interference and the worst-case execution scenario has been experienced when collecting the traces during the second phase.
- It is required to instrument the application source code, which poses the same issue of intrusiveness as in measurement-based approaches. Example tools include Rapitime [11] and MTime [12].

5.1.4 Measurement-based Probabilistic Techniques

With the current hardware designs, the execution time of a given application depends on the states of the hardware components, and those states depend in turn on what has been executed previously. A classic example of such a tight relationship between the application and the underlying hardware architecture is the execution time discrepancy that can be observed when a program executes on a processor equipped with a cache subsystem. During the first execution of the program, every request to fetch instructions and data results in a cache miss and must be loaded from the main memory. At the second execution, this information is already in the cache and need not be reloaded from the memory, which results in an execution time considerably shorter than during the first run. Because of this dependence to past events, the set of measured execution times of the same program cannot be seen as a set of IID (independent and identically distributed) random variables and most statistical tools cannot be applied to analyze the collected execution traces.

The objective of measurement-based probabilistic techniques is to break this dependence on past events, so that one can sample the execution behavior of an application and then derive from the sample probabilistic estimates (of any parameter) that apply to its overall behavior, under all circumstances and in all situations. To achieve this goal, researchers are nowadays working on modifying the hardware components and their arbitration policies to make them behave in a stochastic manner, without losing too much of their performance. For example, by replacing the traditional Least Recently Used (LRU) or Pseudo-LRU (PLRU) cache-replacement policy for a policy that randomly chooses the cache line to be evicted (and assuming that every cache line has the same probability of getting evicted), the time overhead due to cache penalties and cache line evictions can be analyzed as an IID random variable with a known distribution. If every source of interference exhibits a randomized behavior with a known distribution, then the execution time itself can be analyzed statically.

The current trend in probabilistic approaches is to apply results from the extreme value theory (EVT) framework to the WCET estimation problem [12, 13]. In a nutshell, these EVT-based solutions first sample the execution time of an application by running it over multiple sets of input arguments on a randomized architecture that is designed to confer a stochastic behavior on the application runtime. Then, these EVT-based solutions organize the sample into multiple groups/intervals, analyze the distribution of the local maxima within these intervals and then estimate how far the execution time may deviate from the average of that "distribution of the extremes."

Although considerably new, measurement-based probabilistic techniques have been the object of tremendous research efforts in the last few years, most of the breakthroughs in that discipline have been made in the scope of the European projects PROARTIS [14] and PROXIMA [15].

Pros:

- Provide safe and potentially tighter WCET estimates than static and hybrid techniques.
- Provide information not only on the WCET of a program but on the complete spectrum of the distribution of its execution time.

Cons:

- Require modifying the hardware to ensure that the components exhibit a stochastic behavior.
- As the IID requirement is hardly verified in currently available platforms (especially COTS platforms), the applicability of measurement-based probabilistic techniques is limited.

5.2 Our Choice of Methodology for WCET Estimation

As seen in the previous section, there exist several methodologies to estimate the WCET of an application, each with its own advantages and disadvantages. Those methodologies fall into the following main categories, namely static, measurement-based, and hybrid. Here we would like to briefly re-iterate on why among those four methodologies we decided to use a measurement-based approach.

There is currently an evident clash of opinions in the research community about which methodology prevails over the others. During the last two years we had the opportunity to debate with partisans of each of these approaches. It is important to stress that we do not mean to take a side in this book, simply because we recognize that each approach comes with its own set of strengths and weaknesses. Our methodology simply uses the one whose downsides impede as little as possible our objectives. The following subsections summarize our opinion on the matter and present the observations that have driven our choice towards using a measurement-based technique.

5.2.1 Why Not Use Static Approaches?

In this section, we present some of the reasons why we did not choose static approaches to timing analysis, but rather opted for a measurement-based approach. Before going into the details, it is worth mentioning that recent COTS manycore platforms present complex and sophisticated architectures such that it is very challenging at design time, if not impossible, to come up with an accurate model for all the behavioral implications associated with the possible operational decisions that the system can take at runtime. This claim holds true even for the most experienced systems designers.

> *By using hardware platforms such as the Kalray MPPA-256, or any other platform designed to provide high performance, we argue that it is practically infeasible to derive WCET estimates by using static timing analysis techniques.*

In theory, it is always possible to extract safe and reliable timing models and define mathematical abstractions to study the behavior of a deterministic system. However, we argue that it is practically challenging to define and use static mathematical models of the considered platforms, mainly because of:

The inherent system complexity: Typical COTS hardware components are extremely complex. Currently the market of embedded and electronic components is unarguably driven by the ever-increasing need for higher performance. The only way to constantly enhance the performance is to optimize the produced chips and boards by adding all sorts of optimization features. Optimization is achieved by allowing the system to take and revise its operational decisions on-the-fly, at runtime, based on the current workload of the system or any informational data collected about the running application and its environment. Since those decisions are taken at runtime, it is impossible to predict the exact behavior of the system at the analysis time.

The only option for static tools is to assume that the system will most of the time be in a worst-case situation, in which the optimization features will have very little or no effect. This makes static models pessimistic and the produced timing estimates may not reflect accurately the actual timing behavior of the system.

The human resources required: An increased system complexity leads to a longer time-to-model. Developing a draft model of a platform may take up to several years to reach the desired level of accuracy and be validated. Besides this fact, our software stack and methodology aim at being platform agnostic and therefore be applicable to a large set of hardware platforms. To this end, they should provide a generic abstraction between the application logic and the system interfaces so that the development costs and efforts are always reasonable and limited. This is an objective for which the inherent portability of measurement-based solutions appears to be more appropriate.

Portability: The "rigidity" of static approaches: Using static timing analysis techniques goes against our goal of developing a flexible and generic framework which can be "easily" ported to different platforms from various vendors. This has been a key driver in the development of our timing analysis methods, in order to increase the exploitation opportunities in multiple application domains.

The non-availability of the specification details: To devise accurate models, all the information about the target platform must be available and accurate. This is not the case in practice. Chip manufacturer generally keep most information secret, unfortunately.

The complexity of the execution environment: Static timing analysis tools are designed primarily to focus on applications executed *sequentially* in safety-critical embedded systems. Those systems generally provide a very time-predictable and "inflexible" runtime environment in which every mapping and scheduling decision is statically taken at design-time and is then final. Unlike those systems, the software stack considered in this book offers a much more complex and dynamic runtime environment composed of multiple conceptual layers: the code of the RT tasks is executed *in parallel* by being fractioned into OpenMP tasks, those tasks are mapped to clusters, then to threads inside the clusters, and then these threads are scheduled statically or dynamically on the cores. The dynamicity of the processor resource usage ensures a decent application throughput (by maximizing the utilization of the available computing resources) but it naturally impacts adversely on its time-predictability.

> *Traditional hybrid approaches are also not applicable as the complexity of the software stack makes the static control-flow analysis step impossible.*

Since the RT tasks execute in parallel, and even using static mapping approaches, the total order of execution of task-parts is only determined at runtime, it is thus infeasible to investigate all possible scenarios at design-time to identify the worst-case execution flow/path. It is important to re-iterate that traditional timing analysis techniques have been designed primarily to analyze "simple" software codes executed on "simple" and predictable hardware architectures, typically implemented by using low-level programming languages and by obeying strict and specific coding rules to reduce programmer's errors. The framework presented in this book clearly targets much more complex software applications that exhibit a high degree of flexibility and dynamicity in their execution.

5.2.2 Why Use Measurement-based Techniques?

> *In measurement-based approaches, WCET estimations are derived from values that have been observed during the experimentation. What about the values that have not been observed? How can we account for them and be sure that the WCET estimates are reliable?*

Critics of measurement-based approaches for estimating the WCET of an application make a simple yet very valid point. The actual WCET is unknown and is very likely not to be experienced during testing. Even worse, it is not even possible to know whether the worst case has been observed or not. In short, this means that there is no guarantee that such an approach can forecast the exact value of the WCET. All measurement-based techniques implicitly infer a WCET from values for which the "distance" from the actual worst-case is unknown. A direct consequence is that, although those techniques make predictions based on sophisticated and elaborate computations, formally speaking, they can never guarantee that their predictions are 100% "safe". This may be problematic for applications requiring hard RT guarantees, typically in safety-critical systems for instance.

However, one can note that in many application domains, certifiable guarantees based on unquestionable and provable arguments are not required.

For instance, many applications need only "reliable" estimations, in the sense that one must be able to rely on those values and measure the risk of them being wrong (through confidence levels provided by the analysis, for example).

Estimations of the trustworthiness of the produced values (i.e., the confidence in those values) can be expressed through probabilities derived by statistical tools. Specifically, in our approach, the traces of execution times collected at runtime are fed into a statistical framework, called DiagXtrm, in which they are subjected to a set of tests to verify basic statistic hypotheses, such as stationarity, independence, extremal independence, execution patterns/modes, etc. Depending on the results of those tests, it is determined whether the EVT can be applied to those traces. If the tests are successful, the EVT is used to "extrapolate" the recorded execution times and accurately identify the higher values that have not been observed during testing, but for which the likelihood of occurrence is not statistically impossible. Besides this, our framework also provides techniques to assess how "trustworthy" those EVT estimations really are. This last step is of fundamental importance to evaluate the quality of the estimations and find out whether confidence can be placed into the analysis.

Despite all the interesting features provided by the application of EVT to the WCET determination problem, it has been widely criticized in the research community. The main argument against it is that the process of creating the traces (i.e., the execution of an application's code by a given hardware platform) is known to be a process which is neither independent nor identically distributed, which is a prerequisite to the application of the EVT to a data sample. We believe that this argument, although correct because the process is *de facto* not inherently IID, does not allow to conclude on the non-applicability of the EVT. In our view, being an IID process is not necessary, provided that the said process behaves as if it were. This is why the EVT has been applied in so many application domains where it is today recognized to provide helpful and satisfactory results. EVT is used for instance to predict the probability distribution of the amount of large insurance losses, day-to-day market risk, and large wildfires. Needless to say, none of these processes are truly IID.

Whether this is right or not is disputable and we do not intend to close the discussion in this chapter. However, we believe that the doubt this casts on the applicability of the EVT makes this framework worth being investigated further and hopefully will unveil its true potential. In case we are wrong, we will hopefully discover why it is not applicable and close the debate that has been going on already for several years.

> *In measurement-based approaches, the integrity of the actual code*
> *to be deployed in the target hardware is somehow depleted by the*
> *addition of the intrusive instrumentation code to measure the time;*
> *in other words, the measurements themselves add an overhead to*
> *the execution time of the analyzed program.*

This problem can be reduced, e.g., by using hardware measurement tools with no or very small intrusiveness, or by simply letting the added measurement code (and thus the extra execution time) remain in the final program. When doing this, possible disturbances like interrupts also have to be identified and compensated for. The intrusiveness of the instrumentation code is discussed in Section 5.3.5 and we provide efficient solutions to deal with it.

Nearly all the embedded platforms, like the MPPA-256 platform considered in our experimentations, provide a lightweight and non-intrusive trace system that enables the collection of execution traces in predefined time bounds. By using this trace system, we are able to collect meaningful traces of execution without generating too many disturbances in the regular timing behavior of the analyzed application. Based on all the experiments conducted on the Kalray board, we concluded that the time necessary to record a time stamp is 52 clock cycles. By placing "trace-points" (points in the program where the current time is recorded) at well-defined places, we can thus easily subtract the overhead associated with measuring the time itself.

Wrapping things up:

The best candidates for the worst-case timing analysis of the type of workloads considered in this book are the measurement-based approaches. Thus, our proposed methodology relies on timing-related data collected by running the application on the target hardware. This way, we avoid both the burden of modeling the various hardware components (which takes considerable effort and time), as in static timing analysis tools; and the pitfalls and pessimism associated with the over-approximations resulting from the confidentiality, and thus the non-availability, of specific information related to the internal configuration of the components. In addition, the fact that our approach is not tied to specific hardware infrastructures and application designs allows it to benefit from a higher flexibility and portability than static timing analysis methods, and it considerably reduces the time-to-model and time-to-result. In the next sections, we will discuss the specifics of our method and how we propose to overcome or at least mitigate the negative aspects inherent to measurement-based techniques.

5.3 Description of Our Timing Analysis Methodology

5.3.1 Intrinsic vs. Extrinsic Execution Times

The execution time of any piece of code, e.g., a basic block, a software function, or an *OpenMP task-part*, can be seen as composed of two main terms: the *intrinsic* execution time spent executing the instructions of the code, and the *stalling* time, i.e., the time spent waiting for a shared software or hardware resource to become available. To understand how timing analysis is performed in this book, it is fundamental to understand the difference between these two components. If the analyzed software function does not have a functional random behavior (i.e., the outcome of evaluating a condition is never the result of an operation involving randomly generated numbers), then any input dataset always produces one output (and this output remains the same no matter how many times the function is executed on the same input). Further, for a given input dataset, the execution path taken throughout the function's code will always be the same. That is, under this assumption of not involving randomness in the control flow of the analyzed function, running it over a given set of input data over and over again always results in executing the exact same sequence of instructions and eventually, it always produces the same output.

For a given input dataset, we call the "*intrinsic* execution time" of a function the time that it takes to produce its output, assuming that all software and hardware services provided by the execution environment and shared among different cores are always available, and thus the core running that function never stalls waiting for one of these resources to become available. That is, the intrinsic execution time of a function is its execution time when it runs in isolation, i.e., with no interference whatsoever with the rest of the system on the shared resources. On a perfectly predictable hardware architecture where every instruction takes a constant number of cycles to execute, running the same function in isolation over the same set of input arguments should always results in the exact same execution time. Although this may sound like a very strong assumption, we will see that on a platform such as the Kalray MPPA-256 this property is satisfied. By running a preliminary set of tests with the same program an arbitrary number of times over the same inputs, we experienced a variation of its execution time of typically less than 0.1% of the maximum observed.

For a given input dataset, we call the "*extrinsic* execution time" of a function the time that it takes to produce its output, assuming a maximum interference on all the shared resources. That is, the extrinsic execution time of a function is its execution time assuming that all the software and hardware services provided by the execution environment and shared among the cores are constantly saturated by concurrent requests from other system components. Contrary to the intrinsic execution time, on mainstream multicore architectures the extrinsic execution time is subject to huge variabilities due to the high number of processor resources shared amongst software functions.

5.3.2 The Concept of Safety Margins

When testing an application and measuring its execution time, it is very likely, if not certain, that the (usually very rare) situation where the application takes its maximum execution time does not occur. This is due to either of the following reasons:

1. The testing process failed to identify the set of input arguments that takes the longest execution path throughout the program's code, i.e., the path that leads to the WCET.
2. The testing process found the execution path(s) leading to the WCET but did not generate the maximal possible interference while exercising those paths. This means that the actual WCET is not observed only because the interference patterns generated during testing did not put the application into the worst execution conditions.

Regarding the first case, for most programs, the number of possible execution paths (in comparison to the high number of possible inputs) is too large to make exhaustive testing possible and/or realistic. Therefore, measurements are carried out only for a subset of input values. Typically, the testing process starts with the identification of a set of potentially "nasty" inputs that are likely to make the program take the longest execution path throughout its code and provoke its WCET. This step is typically supervised and based on some manual inspection of the code. Note that powerful tools exist such as the Rapita Verification Suite (RVS) that incorporates a code-coverage tool (RapiCover [16]) to test all parts of a given code and guarantee its full coverage during testing. We believe that such tools may be employed to help system designers identify the "worst" input datasets.

The problem of defining the worst input dataset(s) is thus not new, and to some extent it is independent of the underlying hardware architecture. Of course, the execution time of a given path depends on the execution time of each instruction in that path, and therefore is dependent on the architecture, but the method to search the space of all possible inputs and identify those that lead to the longest execution path is platform-agnostic. Since the problem was already there on single-core architectures, with mature solutions for it, we do not focus, in this book, on improving this part of the process.

Regarding the second point, it is always assumed that the worst-case interference is not observed during testing and therefore the maximum execution time recorded is an under-approximation of the actual WCET. To compensate for this, it is common to add a safety margin to the measured WCET, in the hope that the actual WCET lies below the resulting augmented estimation. The main question that remains open is whether the extra safety margin provably provides a safe bound, since it is based on some informed estimates. In principle, a very high margin yields an upper-bound on the execution time that is likely to be safe (i.e., greater than the actual WCET), but results in an over-dimensioned system with a low utilization of its resources, whereas a small margin may lead to an under-estimation of the actual system (worst-case) needs.

Traditionally, the magnitude of the safety margin applied to the maximum measured execution time is based on an estimation of the maximum interference (from the system or from other applications) that has not been observed during the testing phase but that the analyzed application could potentially incur at runtime. For single-core systems, this estimation of the worst-case interference is usually built on past experience. For example, in the IEC 61508 standard [17] related to functional safety of electrical/electronic/ programmable electronic safety-related systems, to ensure that the working capacity of the system is sufficient to meet the specified requirements, it is mentioned that:

> "For simple systems an analytic solution may be sufficient, while for more complex systems some form of simulation may be more appropriate to obtain accurate results. Before detailed modeling, a simpler 'resource budget' check can be used which sums the resources requirements of all the processes. If the requirements exceed designed system capacity, the design is infeasible. Even if the design passes this check, performance modeling may show that excessive delays and response times occur due to resource starvation. To avoid this situation, engineers often design systems to use some fraction (for example 50%) of the total resources so that the probability of resource starvation is reduced."

As explained above, it is a common practice to simply add a margin of 50% (or any other percentage depending on the user's preferences and his level of confidence in those margins) to the maximum execution time observed. Unfortunately, on multicore and manycore architectures, experts are not yet able to safely estimate reliable margins, as there is no prior experience to be relied upon. Hence, we must build a new body of knowledge and investigate novel approaches to produce reliable timing estimates and margins, and we must motivate these estimations and justify why we believe they are reliable. Our move towards this ambitious goal is described in short in the following subsection.

5.3.3 Our Proposed Timing Methodology at a Glance

In this book, we devised methods to extract both the intrinsic and extrinsic execution times. The overall timing analysis methodology consists of four steps:

Step 1: Extraction of the maximum intrinsic execution time

To measure the maximum intrinsic execution time (MIET), we run the analyzed task sequentially on one core and we configure the execution environment in such a way that no other tasks can interfere with its execution. That is, everything is done to nullify the interference with other applications or with the system itself. This way we put the analyzed task in "ideal" execution conditions in which, in the absence of interference, the time taken to execute its code can be assumed to be due solely to the execution of its instructions (without any stalling time). In these conditions, the task to be analyzed is run multiple times, non-preemptively, over a finite set of input data. These input data have been pre-selected and identified as particularly "nasty", i.e., very likely to make the task take its longest execution path throughout its code and provoke its WCET. We do not elaborate on how to select those inputs.

Step 2: Extraction of the maximum extrinsic execution time

The maximum extrinsic execution time (MEET), on the contrary, is obtained by measuring the time taken to execute the analyzed task in conditions of "extreme" interference. That is, everything is done to maximize the interference with other applications and with the system itself. Measuring the execution time of the analyzed task in those "worst" conditions and over the "worst" input datasets give an estimation of the maximum execution time that the task may experience in the presence of other tasks running concurrently.

Step 3: Extract the execution time after deployment

The MIET and MEET can be considered as lower and upper bounds on the actual WCET of the analyzed task, since they estimate the WCET in conditions of no and extreme interference, respectively. These two estimations are useful to the system designers to understand the impact that tasks may have on each other's timing behavior. For instance, it may be desirable to derive a static mapping of the task-parts to the cores in which the task-parts (the portions of code for which the executions are timed or measured) that are highly sensitive to interference (i.e., the difference between their MEET and MIET is large) are mapped to specific cores in a way that they cannot interfere with each other at runtime.

After taking mapping and scheduling decisions based on the values of the MIET and MEET, these decisions are implemented and the whole system is run in its final configuration. Measures are taken again, this time to estimate the execution time of the tasks in its "final" execution environment, i.e., the environment corresponding to the "after-deployment". Timed traces are recorded like in the previous step and are passed to step 4.

Step 4: Estimate a worst-case execution time

The traces collected in Step 3 reflect the actual execution time of every task-part, and from those their individual WCET can be derived or estimated. The simplest way to proceed is to retain the maximum execution time observed as the actual WCET. For safety purpose, an arbitrary extra "safety margin" can be added to that WCET estimation to make it even safer. The magnitude of the margin depends on how much "safer" the system designers want to be, but we would recommend using a margin that does not exceed the MEETs of the tasks (because the MEETs represent the WCET of the tasks in execution conditions that are unlikely to happen at runtime).

However, instead of arbitrarily choosing a margin, we advocate the use of statistical methods to analyze the traces and make a more "educated" choice driven by mathematical assumptions and computations rather than just a "gut feeling". In this book, we use DiagXtrm, a complete framework to analyze timed traces and derive pWCET estimates.

In the next subsections, we describe every step of our methodology.

5.3.4 Overview of the Application Structure

Before we go to the details, let us briefly recall the type of workloads that we are handling in this book and recap what exactly needs to be measured.

In the considered system model, the application comprises all the software parts of the systems that operate at the user-level and that have been explicitly defined by the user. The application is the software implementation (i.e., the code) of the functionality that the system must deliver to the end-user. It is organized as a collection of RT tasks.

An RT task is a recurrent activity that is a part of the overall system functionality to be delivered to the end-user. Every RT task is implemented and rendered parallelizable using OpenMP 4.5, which supports very sophisticated types of dynamic, fine-grained, and irregular parallelisms.

An RT task is characterized by a software procedure that must carry out a specific operation such as processing data, computing a specific value, sampling a sensor, etc. It is also characterized by a few (user-defined or computed) parameters related to its timing behavior such as its WCET, its period, and its deadline. Every RT task comprises a collection of task regions whose inter-dependencies are captured and modeled by a directed acyclic graph, or DAG.

A task region is defined at runtime by the syntactic boundaries of an OpenMP task construct. For example:

```
#pragma omp task
{
        // The brackets identify the boundaries of the task region
}
```

Hence, hereafter we refer to task regions as *OpenMP tasks*. The OpenMP tasking and acceleration models are described in detail in Chapter 3.

An *OpenMP task-part* (or simply, a task-part) is a non-preemptible portion of an OpenMP task. Specifically, consecutive task scheduling points (TSP) such as the beginning/end of a task construct, the synchronization directives, etc., identify the boundaries of an OpenMP task-part. In the plain OpenMP task scheduler, a running OpenMP task can be suspended at each TSP (not between any two TSPs), and the thread previously running that OpenMP task can be re-scheduled to a different OpenMP task (subject to the task scheduling constraints).

The DAG of task regions can therefore be further expanded to form a typically bigger DAG of task-parts. This new graph of task-parts is called the extended task dependency graph (eTDG) of the RT task. Figure 5.2 shows the eTDG of an example application. Our objective is to annotate every node, i.e., task-part, of the eTDG with an estimation of its WCET and then perform a schedulability analysis of the entire graph to verify that all the end-to-end timing requirements were met.

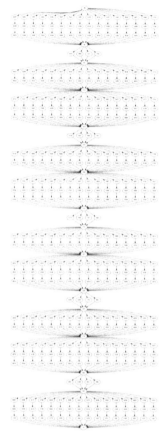

Figure 5.2   Extended task dependency graph (eTDG) of an example application.

5.3.5 Automatic Insertion and Removal of the Trace-points

In this subsection, we discuss how to respectively *insert* (Subsection 5.3.5.1) and *remove* (Subsection 5.3.5.2) trace-points in a given program in an automatic manner.

5.3.5.1 How to insert the trace-points

To measure the execution time of a task-part, we insert a trace-point at its entry and exit points. A trace-point is a call to a system function that records the current timestamp. Therefore, the system will record the time of entering the task-part (i.e., when its execution starts) and the time at which it exits it; the difference between the two straightforwardly gives the time spent executing the task-part.

Inserting the trace-points into the tasks' code can easily be done by the compiler itself, when creating the executable file. Moreover, upon compiling the code and creating the TDG, the compiler can assign a unique Identifier (ID) to every task-part. Overall, this ID can be used to define a trace-point for the task-part associated with an execution time. For example, using the trace system from the Kalray SDK, we ask the compiler to add the following trace-points at the beginning and end of every task-part as illustrated in the code snippet below:

```
#pragma omp task
{
// The brackets identify the boundaries of the task region
        mppa_tracepoint ( psocrates , taskpartID__in ) ;
        /* code of the task-part */
        mppa_tracepoint ( psocrates , taskpartID__out ) ;
}
```

These trace-points indicate to the Kalray MPPA runtime environment that a time-stamp must be recorded each time the execution meets one of these points (together with the ID of the corresponding task-part). The first argument (here, "psocrates") is the name of the "trace-point provider". The user defines it to help him organize all its trace-points into groups. Informally, it can be thought of as a folder name. The second argument is the name of the trace-point. For every task-part we insert a trace-point called "taskpartID__in" at the beginning of the task-part and another trace-point called "taskpartID__out" at the end. We do so because the objective of our next tool is to find every matching pair "*__in/*__out" of trace-points and compute the difference of timestamps (which naturally corresponds to the execution time of the task-part).

Once all the trace-points are correctly placed into the source code, the compiler must create a separate header file "*tracepoints.h*" in which all the trace-points are declared and then include that file in all source files in which trace-points are used (#include "tracepoints.h").

```
#ifndef _TRACEPOINTS_H_
#define _TRACEPOINTS_H_
#include "mppa_trace.h"

MPPA_DECLARE_TRACEPOINT(psocrates, taskpartID__in,())
MPPA_DECLARE_TRACEPOINT(psocrates, taskpartID__out, ())

... // more trace-points

#endif
```

5.3.5.2 How to remove the trace-points

After the analysis step, when the system is ready to be deployed, it is preferable to remove all the trace-points in order not to leave some "dead code." A code is said to be dead either if it is never executed, or when its execution does not serve any purpose, like for example taking time-stamps and not recording them into a file (which would happen if those trace-points were to be left in the source code when compiling the application to be deployed). However, removing trace-points is not a benign operation.

To illustrate the problem that may arise from removing the trace-points, let us consider the following code.

```
int run_index;
for ( run_index = 0 ; run_index < NB_RUNS ; run_index++ ) {
        mppa_tracepoint(psocrates, main__in);
        user_main();
        mppa_tracepoint(psocrates, main__out);
}
```

The `user_main()` function is a call to the main function of the benchmark program "statemate.c" provided by (15). If we disable all compiler optimizations during the compilation phase (this is important and will play a role later) and run this code 100 times on a single core of a compute cluster of the Kalray MPPA-256, we observe that the execution time oscillates consistently between 88492 and 88497 cycles (see Figure 5.3, left-hand side).

Now, let us add to that code a variable *x* to which we assign an arbitrarily chosen integer (here, 1587) as shown below:

```
int x = 1587;
int run_index;
for ( run_index = 0 ; run_index < NB_RUNS ; run_index++ ) {
        mppa_tracepoint(psocrates, main__in);
        user_main();
        mppa_tracepoint(psocrates, main__out);
}
```

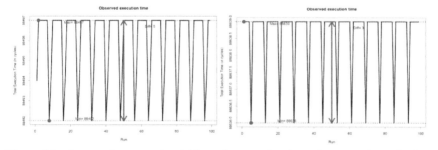

Figure 5.3 Impact of an unused variable on the execution time of an example application.

It is important to stress that the variable x is never used in the program. Since all compiler optimizations are disabled, the variable is not removed from the code by the compiler and is present in the assembly code that it produces. As seen in the Figure 5.3 (right-hand side), the execution time now oscillates consistently between 88, 639 and 88, 636 cycles. This means that the addition of an unused variable to a part of the code which is not even under analysis adds around 140 cycles to the execution time of the measured portion of the code.

This increase in the execution time stems from the fact that after the addition of the line "int x = 1587" to the source code, all subsequent instructions got offset in the system memory by two times the length of an instruction, i.e., the line "int x = 1587" translates to two assembly instructions: one for allocating memory to the variable x and another one for moving the constant "1587" into it. Therefore, the portion of the code being timed has a different "memory layout" as it is mapped to the system memory two "instruction-lengths" further. This in turn impacts on the way the instructions of that part of the code are mapped at runtime to the instruction cache lines and ultimately it results in a perceptive difference in the execution time.

A consequence of this phenomenon is that removing the trace-points after the analysis phase may have for effect to substantially, or at least noticeably, alter the timing behavior of the application and all its task-parts. We came up with two potential solutions to this problem. The simpler one is to leave the trace-points in the code when compiling it for the final release of the application. Although it is a suitable work-around to the memory-shift problem described above, most designers are not in favor of having a dead portion of code, as explained above.

Our second solution is to measure the length, in number of assembly instructions, of the code being executed each time the function `mppa_tracepoint (... )` is called and replace every such call with an equivalent number of NOPs (No Operation assembly instruction). This way neither the semantic of the code nor the memory layout are altered when removing the trace-points. We believe this solution to be both feasible and suitable for use in industrial applications.

5.3.6 Extract the Intrinsic Execution Time: The Isolation Mode

In order to extract the MIET of a task-part, we must start its execution and make sure that it is isolated from the rest of the system. That is, we must nullify all external interference by turning off every other component that could potentially interfere with (and hence delay) the execution of the analyzed task-part. This is achieved by assigning every task-part of the analyzed real-time task to the same thread, and thus to the same core of the same cluster, and then making sure that all the other cores are kept idle. In other words, under this configuration, the RT task is executed sequentially in a single core. However, the intention of this phase is to analyze the execution time of each task-part in isolation, i.e., without suffering interferences, and not the overall RT task execution time. We call this configuration the *isolation mode*; the real-time task is then said to run *in isolation*.

To setup and enforce this isolation mode, we have implemented a platform-specific API. The current version has been written for the Kalray MPPA-256. The API provides a set of easy-to-use functions to configure the execution environment, as well as a set of global parameters and functions that are used to make sure that:

1. all the openMP tasks are assigned to a single thread,
2. the IO cores and the cluster cores are in sync so that the environment is "sanitized" before and after the execution of every openMP task (nothing runs in the background that could interfere with the execution of the analyzed task), and
3. additional functions allow the user to perform specific operations, either before the runtime, such as deciding the memory-mapping and cache-management policy, or during the runtime, such as invalidating the instruction or data caches before executing each task-part.

The main objective of the API is to create a controlled environment in which every task-part is run over a specific set of inputs and is isolated from the rest of the system so that it incurs minimum interference during its execution.

5.3.7 Extract the Extrinsic Execution Time: The Contention Mode

To extract the MEET of a task-part, we start the task and interfere as much as possible with its execution at runtime. The objective of the contention mode is to create the "worst" execution conditions for the task-parts so that their execution is constantly suspended due to interference with other tasks. In this step, for each task-part, we record the maximum execution time observed under those conditions. This gives us an estimation of the maximum execution time of each task-part when it suffers interference from other tasks on the shared resources.

This *contention mode* is similar to the isolation mode in that all the task-parts of the analyzed real-time task are assigned to the same thread, and thus to the same core within a same cluster, effectively executing the RT task sequentially. However, contrary to the isolation mode that shuts down all the other cores of the cluster (thereby nullifying all possible interference within that cluster), we deploy onto all these other cores small programs called IG, which stands for *Interference Generator*. Those programs are essentially tiny pieces of code that have the sole purpose of saturating all the resources (e.g., interconnection, memory banks) that are shared with the task-parts under analysis. Recall that the objective of the contention mode is to create the worst execution conditions for the execution of the task-parts, conditions in which their execution is slowed down as much as possible due to contention for shared resources.

Implementing the IG that generates the worst possible interference that a task-part could ever suffer is a very challenging, if not impossible, task. This is because the exact behavior of the task-part to be interfered with (i.e., its utilization pattern of every shared resources and the exact time-instants of accessing it) should be known, as well as all the detailed specifications of the platform. Besides, even if that information was known, the execution scenario causing the maximum interference may be impossible to reproduce. Rather than concentrating our efforts on creating such a "worst IG", we have opted for the implementation of an IG that is "bad enough" and used it as a proof of concept to demonstrate how large the time-overhead incurred by the task-parts due to the interference can be.

Our implementation of the IG consists of a single function *IG_main* that is executed by a thread dispatched to every core on which the task-parts are not assigned (recall that the application under analysis is executed sequentially in a single core). That is, every core that is not running the task-parts runs a thread that executes *IG_main*. Essentially, *IG_main* executes three functions, namely:

1. IG_init_inteference_process ()
2. IG_generate_interference ()
3. IG_exit_inteference_process ()

The first one is called upon deploying the IG, at the beginning of *IG_main*, before the task-parts start to execute and be timed. The second one is the main function. It creates interference on the shared resources. The call to that function is encapsulated in a loop that terminates only when the *IG_main* is

```
int* my_array;
inline void IG_init_interference_process()
__attribute__((always_inline));
inline void IG_generate_interference()
__attribute__((always_inline));
inline void IG_exit_interference_process()
__attribute__((always_inline));
```

explicitly told to stop. Finally, the third function is called when all the task-parts have been timed and the analysis process is about to end.

Let us now briefly describe our implementation of the IG on the Kalray MPPA-256. This implementation is provided in a single file, which starts with the declaration of an array of integer called `my_array` and declares the three main functions as described above. The `__attribute__((always_inline))` instruction is used to enforce and oblige the compiler to use inlining for these three methods. The inlining technique is used to waste as little time as possible jumping from one address to another in the code, as jumping does not create interference.

Below is a code snippet of the first function "*IG_init_interference_process()*."

```
inline void IG_init_interference_process() {
        int array_size = 1024;
        // Create an array of Integers. One integer is 4 bytes
        my_array = malloc(array_size * sizeof(int));
        // Fill the array with numbers.
        int cpt = 0;
        for (cpt = 0 ; cpt < array_size ; cpt++) {
                my_array[cpt] = cpt;
        }
}
```

This function simply allocates memory to `my_array` (1024 integers) and fills that memory space with arbitrary values. Note that on the Kalray MPPA-256, a thousand integers occupy roughly half of the private data cache of a VLIW[2] core in a compute cluster.

The third function, "*IG_exit_inteference_process()*", is the simplest as it only frees the memory space held by `my_array` as shown below.

```
inline void IG_exit_interference_process() {
        Free(my_array);
}
```

The second function, "*IG_generate_interference ()*," is the main one and a snippet of its code is presented below.

---
[2] Very Long Instruction Word.

```
inline void IG_generate_interference() {
        __builtin_k1_dinval();
        __builtin_k1_iinval();
            register int *p = my_array;
        volatile register int var_read;
        var_read = __builtin_k1_lwu(p[0]);
        var_read = __builtin_k1_lwu(p[8]);
        var_read = __builtin_k1_lwu(p[16]);
        var_read = __builtin_k1_lwu(p[24]);
        var_read = __builtin_k1_lwu(p[32]);
        var_read = __builtin_k1_lwu(p[40]);
        var_read = __builtin_k1_lwu(p[48]);
        var_read = __builtin_k1_lwu(p[56]);
        var_read = __builtin_k1_lwu(p[64]);
        var_read = __builtin_k1_lwu(p[72]);
        var_read = __builtin_k1_lwu(p[80]);
        (...)
        var_read = __builtin_k1_lwu(p[1007]);
        var_read = __builtin_k1_lwu(p[1015]);
        var_read = __builtin_k1_lwu(p[1023]);
}
```

The function starts by invalidating the content of the data and instruction caches. Then, it reads every element of "*my_array*", starting from the element K = 0 and moving on iteratively from element K to element ((K+8) mod 1024), until K reaches 1023. This way, every element of the array is read exactly once and every two consecutive readings access data that are located exactly 8 * 4 = 32 bytes apart in the memory (the size of an integer is standard on the Kalray, i.e., 4 bytes). This is done on purpose knowing that the private data cache line of every VLIW core in the compute clusters of the Kalray MPPA-256 is 32 bytes long. Consequently, every reading causes a cache miss and the value must then be fetched from the 2 MB in-cluster shared memory, hence it creates traffic on the shared memory communication channels and potentially interferes with the task-part being analyzed.

By running the task-parts concurrently with these IGs, every request sent by a task-part to read or write a data in the shared memory is very likely to interfere with a read request from one of the IGs. We have conducted experiments on the Kalray MPPA-256 using several use-case applications to evaluate the magnitude of the increase in the execution time due to this interference. Depending on the configuration of the board and the memory footprint of the task-parts and their communication pattern with the memory, the difference between the maximum execution time observed in isolation mode and in contention mode is substantial as the execution time of a task-part may be increased by a factor of 9.

5.3.8 Extract the Execution Time in Real Situation: The Deployment Mode

After determining the intrinsic and extrinsic execution times (i.e., the MIET and the MEET), we communicate them to the mapping and scheduling analysis tools through the annotation of the TDG of the real-time task. Once all necessary mapping and scheduling decisions are taken, the application is run again, but this time in its final production environment. This means that the platform configuration and mapping and scheduling decisions are no longer imposed and defined so as to create specific execution conditions. Then, we collect runtime-timed traces of the task-parts in their final environment, without any supervision or any attempt to explicitly favor or curb the execution of the application.

5.3.9 Derive WCET Estimates

As already discussed, the traces collected in the previous step reflect the actual execution time of every task-part when they run in their final environment, under different execution conditions. The objective of this final step is to derive WCET estimates from those traces. The simplest solution is to retain the maximum execution time observed during the deployment mode as the actual WCET and, for safety purposes, add an arbitrary "safety margin" to that maximum to make it "safer". The magnitude of the margin depends on how much "safer" the system designers want to be, but we would recommend using a margin that does not exceed the MEET. However, instead of arbitrarily choosing a margin, we advocate the use of statistical methods to analyze the traces and make a better thought out choice.

The objective of Measurement-Based Probabilistic Timing Analysis (MBPTA) approaches is to characterize the variability in the execution time of a program through probability distributions and in particular, they aim at deriving probabilistic WCET estimates, a.k.a. pWCET. A pWCET is a probability distribution of the WCET of a program. That is, through MBPTA, the WCET is no longer expressed as a single value but as a range of values, each assigned to a given probability of occurrence with the obvious relation: the higher the value assumed to be the WCET, the lower its probability of occurrence. Based on this framework system designers are in a position to somewhat decide on the reliability of the final WCET estimation, simply by ignoring all values for which the probability of observing an execution time greater than those exceeds a pre-decided threshold. The EVT is a popular theoretical tool used by most MBPTA approaches. The EVT aims at modeling and estimating better the tail of a statistical distribution, which is *de facto* what the MBPTA is trying to achieve when focusing on the pWCET.

Researchers at the French Aerospace Lab (ONERA), in France, recently proposed a remarkable framework and tool to analyze timed traces and derive pWCET estimates. The framework is called DiagXtrm [18] and defines a methodology composed of three main steps:

1. Analyze the traces
2. Derive pWCET estimates using the EVT
3. Assess the quality of the estimations.

Together with the theory and the definition of the methodology, they developed a tool to diagnose execution time traces and derive safe pWCET estimates using the EVT. However, the EVT can be applied to a given trace only if some hypotheses are verified. Testing those hypotheses is the focus of the first step ("Analysis of the traces") above.

In a nutshell, for safely applying the EVT and getting reliable pWCET estimates, one has to check a few hypotheses including for instance stationarity, short-range dependence, and extreme independence. The *stationarity* of a trace reveals whether measurements belong to the same probabilistic law without knowing it. The *independence* (short-ranged or between the extremes) analysis aims at determining whether there are obvious correlations within the measurements. Systemic effects in a modern hardware platform are so complex and numerous that it is quite impossible to infer the probability of happening of an execution time knowing the value of the preceding ones, i.e., the execution time of an application cannot be inferred from the execution times of its previous executions. System non-determinism, coming from the considered system's degree of abstraction, knowledge, and randomness observed in a timed trace motivate the independence of the measurements that has to be studied at "different scales" (i.e., short-range independences and independences of the extremes). DiagXtrm implements the most advanced tests to verify the stationarity hypothesis and measure the degree of correlation between patterns of different lengths within a trace. Thus, it studies both short-range and distant dependencies between the measurements.

If all the hypotheses are verified, then the EVT is applied to produce pWCET estimates. These estimates are the result of sophisticated computations based on parameters that must be carefully set. The user is in charge of setting those parameters as he wants, and thus has a great influence on the pWCET estimation process. Note however that the DiagXtrm tool provides helpful functions to guide the choice of many of those input parameters.

Finally, the tool features a set of tests to evaluate the quality of the produced estimates, together with other tests to assess the confidence that all the hypotheses were verified. We believe that this last phase is fundamental and is a first step towards building confidence and assessing the reliability of the pWCET estimates.

5.4 Summary

The analysis of the timing behavior of software applications that expose real-time requirements and dedicated to execute on the recent COTS manycore platforms such as the Kalray MPPA-256 raises a number of important issues. Because a reliable and tight WCET estimation for each task running on such a platform is a crucial input at the schedulability analysis level, we showed that it is neither acceptable nor realistic to ignore all the interactions between each analyzed task, the OS, and all the other tasks running in the system. Then, depending of the type of workload that is considered, we also showed that the choice of the methodology to be adopted must be conducted with care. In this chapter, after presenting an overview of all the possible methodologies, and after discussing their advantages and disadvantages, we opted for a measurement-based approach. We explained and motivated this choice and finally presented the details of our solution. Here, we showed that both the intrinsic (MIET) and extrinsic (MEET) execution times of each task are pivotal values to be extracted in order to guide the designer in deriving a reliable and tight WCET.

References

[1] *OTAWA*. Available at: http://www.irit.fr/recherches/ARCHI/MARCH/OTAWA/doku.php?id=doc:computing_a_wcet.
[2] Ermedahl, A., Engblom, J., "Execution Time Analysis for Embedded Real-Time Systems," eds. Joseph, Y-T., Leung, S. H., Son, I. L., Chapman and Hall/CRC – Taylor and Francis Group, 2007.
[3] Lokuciejewski, P., Marwedel, P., *Worst-Case Execution Time Aware Compilation Techniques for Real-Time Systems – Summary and Future Work* (Springer: Netherlands), pp. 229–234, 2011.
[4] *AbsInt GmbH*. Available at: http://www.absint.com/ait/analysis.htm.
[5] *Tidorum Ltd*. Available at: http://www.bound-t.com/.
[6] *NUS*. Available at: http://www.comp.nus.edu.sg/~rpembed/chronos/.

[7] *IRISA.* Available at: http://www.irisa.fr/alf/index.php?option=com_content&view=article&id=29&Itemid=&lang=fr.

[8] *MRTC.* Available at: http://www.mrtc.mdh.se/projects/wcet/sweet/DocBook/out/webhelp/index_frames.html.

[9] Kirner, R., Puschner, P., Wenzel, I., "Measurement-based worst-case execution time analysis using automatic test-data generation." *4th Euromicro International Workshop on WCET Analysis*, pp. 67–70, 2004.

[10] *Rapita Systems Ltd.* Available at: http://www.rapitasystems.com/products/rapitime/how-does-rapitime-work.

[11] Carnevali, L., Melani, A., Santinelli, L., Lipari, G., "Probabilistic Deadline Miss Analysis of Real-Time Systems Using Regenerative Transient Analysis." In *Proceedings of the 22nd International Conference on Real-Time Networks and Systems*, Versaille, pp. 299–308, 2014.

[12] Santinelli, L., Morio, J., Dufour, G., Jacquemart, D., "On the Sustainability of the Extreme Value Theory for WCET Estimation." 14th International Workshop on Worst-Case Execution Time Analysis, Versailles, pp. 21–30, 2014.

[13] *Proartis: Probabilistically Analysable Real-Time Systems.* Available at: http://www.proartis-project.eu/.

[14] *Probabilistic real-time control of mixed-criticality multicore and many-core systems (PROXIMA).* Available at: http://www.proxima-project.eu/.

[15] *Rapita Systems Ltd.* Available at: https://www.rapitasystems.com/products/rapicover.

[16] *The International Electrotechnical Commission. Functional Safety of Electrical/Electronic/Programmable Electronic Safety-Related Systems – Part 7, 2nd Edition, Requirement C.5.20 (Performance Modeling)*, Geneva, p. 99, 2010. IEC 61508.

[17] Gustafsson, J., Betts, A., Ermedahl, A., Lisper, B., "The Mälardalen WCET benchmarks – past, present and future." In *Proceedings of the 10th International Workshop on Worst-Case Execution Time Analysis (WCET'2010)* Brussels, Belgium, pp. 137–147, 2010.

[18] Onera. *Onera – DiagXTrm*, Available at: https://forge.onera.fr/projects/diagxtrm2.

[19] MTime, *Vienna real-time systems group*, Available at: http://www.vmars.tuwientuwien.ac.at.

6

OpenMP Runtime

Andrea Marongiu[1,2], Giuseppe Tagliavini[2] and Eduardo Quiñones[3]

[1]Swiss Federal Institute of Technology in Zürich (ETHZ), Switzerland
[2]University of Bologna, Italy
[3]Barcelona Supercomputing Center, Spain

This chapter introduces the design of the OpenMP runtime and its key components, the *offloading library* and the *tasking runtime library*. Starting from the execution model introduced in the previous chapters, we first abstractly describe the main interactions among the main actors involved in program execution. Then we focus on the optimized design of the offloading library and the tasking runtime library, followed by their performance characterization.

6.1 Introduction

The model assumed in the previous chapters considers the existence of multiple applications, starting execution on the *host* processor, and each one is composed of multiple real-time (RT) tasks which can be sent to the *accelerator* with the aim to speed up their execution. This paradigm, commonly referred to as *offloading*, has been widely adopted in many computing domains from embedded systems to HPC [1, 2]. In the context of the Kalray architecture (described in Chapter 2), IO cores take on the *host* role while the *clusters* are used as accelerators. Accordingly, an OpenMP-based software stack with offloading support must leverage both host and acceleration roles. On the *host* side, an OpenMP directive (#pragma omp target) is used to specify a region of code which can be offloaded. Inside a cluster, a pool of threads is dedicated for the execution of the offloaded workload and the RTOS (introduced in Chapter 7) is in charge of scheduling the execution of the threads on the available cores.

The complete software stack to handle the described execution model is composed of an *offloading library* and a *tasking runtime library*. The offloading library executes on the *host* and is in charge of initiating offload sequences to the accelerator. On the accelerator side, the *request manager* (RM) is the component in charge to collect offload requests and create pools of threads (hereafter called *jobs* as in the OS terminology) to execute them. Depending on runtime design and hardware specific features, the RM can be implemented as an RT task (a software component) or may be mapped to a dedicated core (a hardware component). The tasking runtime library provides an optimized support for task parallelism on the accelerator and runs on top of the RTOS. It is further divided into a low-level library (or LL-RTE) [3], where all the tightly coupled interactions with the RTOS are implemented, plus a high-level library, where all the management of the tasking constructs resides.

6.2 Offloading Library Design

Figure 6.1 summarizes the timing diagram (time flows from top to bottom on the vertical axis) and the interactions between the software blocks providing the offload support. At the higher level of abstraction, the *host* sends request to the RM, which orchestrates the execution of the workload on the *processing elements* (PEs).

The *host* support is implemented as a user-level library that interfaces OpenMP offloads (expressed at the application level with a `target` directive) to the computing clusters. The key features of this library can be summarized as follows:

- **Low-cost offload:** As initializing the communication channels between the *host* and the offload manager and loading into the cluster shared memory the binary file containing the OpenMP library (high-level library + LL-RTE) are costly operations, the *host* offload library implements it as a one-time operation that happens at system startup (the `GOMP_init` method). Every time the *host* program encounters a `target` directive, this is translated into a call to the `GOMP_target` function, which sends a control packet to the offload manager of the target cluster and then triggers the copy of input data. This handshake procedure is streamlined to guarantee minimum overhead.
- **Asynchronous offload:** The offload procedure is asynchronous. After sending the offload request to the cluster, `GOMP_target` immediately

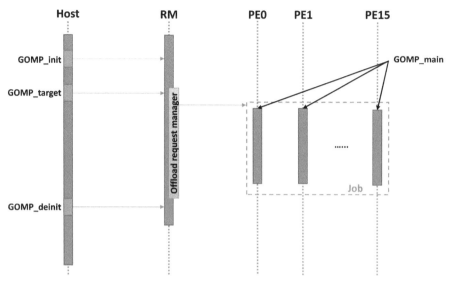

Figure 6.1   Timing diagram of an offloading procedure.

returns to the caller (with the exception of the multi-offload case described below). The result of the offload computation can be retrieved by calling a blocking synchronization primitive (GOMP_target_wait).

- **Multi-cluster support:** An application can perform offloads on different clusters, from 1 up to 16. The initialization is required for each cluster that is used by the current application. The cluster is specified by the programmer using the OpenMP syntax (i.e., the device clause of a target directive).
- **Multi-offload support:** An application can perform multiple offloads on the same cluster. At the same time, multiple offloads can coexist on the same cluster at different priority levels[1]. The priority level is specified by the programmer using the OpenMP syntax (i.e., the priority clause), and it is propagated to the runtime using a parameter of GOMP_target.

The function calls to the offloading runtime are not invoked directly by the developer, as the OpenMP syntax is used to identify the code and data to be offloaded. The compiler transforms the offloading OpenMP directives as

---

[1]Focusing on the MPPA-256 platform, currently two levels are supported on RTEMS hosts and four on Linux hosts

defined in its accelerator model (i.e., target and declare target directives) to the corresponding offloading runtime calls, as described in Chapter 3.

In our design, the RM is implemented as a persistent RTOS task to be executed on the accelerator side (i.e., by one of the cluster cores). The RTOS leverages the notion of a *task scheduling point* (TSP) to check the availability of a new offload request and perform the requested actions. At each such scheduling point, the RTOS can (re)start the execution of the RM itself (if a new offload has arrived in the meantime and needs to be enqueued to the ready job list) or another job in the queue, depending on the synchronization policy adopted. TSPs are naturally identified as synchronization points in an OpenMP program (see Chapter 3 for more details). The OS provides synchronization primitives (described in Chapter 7) which can be used to block one (or more) thread(s) within a job on a certain wait condition, and the OpenMP runtime invokes these primitives to enforce synchronization.

To reduce the runtime overheads, the metadata for all the supported RTOS jobs on a cluster (one per priority level) are created and initialized upon the first call to the RM. The activated jobs execute the GOMP_main function of the runtime library to initialize the offload support on the cluster side.

6.3 Tasking Runtime

The OpenMP tasking model has been introduced in Chapter 3. Task-based parallelism offers a powerful conceptual framework to exploit irregular parallelism in target applications, and several works have demonstrated the effectiveness of tasking [4–7]. However, the sophisticated semantics of the OpenMP tasking execution model are translated into a complex control code that has to be executed in addition to the application code itself. This ultimately results in significant time overheads, if the application tasks are not large enough to hide such overheads. Thus, a performance-efficient design of a tasking runtime environment (RTE) targeting low-end embedded manycore accelerators is a challenging task, as embedded parallel applications typically exhibit very fine-grained parallelism [6, 8], and are thus very sensitive to time overheads. Moreover, memory overheads are also very relevant in this context, as embedded architectures feature very limited amounts of fast, on-chip memory. Allocating runtime support metadata in such memories reduces time overheads, as the control code executes faster, but reduces the space available for program data. As the metadata for a tasking runtime might consume a significant amount of memory, it is necessary to find a good tradeoff between the implied space and time overheads.

The applicability of the tasking approach to embedded applications and embedded manycore accelerators is often limited to coarse-grained parallel tasks, capable of tolerating the high overheads typically implied in a tasking runtime. State-of-the-art tasking runtimes for embedded manycores [6] succeed in achieving low overheads and enabling high speedups for very fine-grained tasks, but only for simple *flat* parallel patterns (i.e., where all the tasks are created from the same parent task). The main reason for this limitation lies in a key design choice: only *tied* tasks are supported by most RTEs, whereas *untied* tasks are not supported. If a *tied* task is suspended (due to synchronization, creation of another task, etc.), only the thread that initially owned it is allowed to resume its execution. This clearly significantly limits the available parallelism when more sophisticated (and realistic) parallel execution patterns are considered, like nested tasking (for instance, in programs that use recursion).

Scheduling policies: Another limitation that follows from supporting only *tied* tasks is the restricted set of scheduling policies available. Breadth-first scheduling (BFS) and work-first scheduling (WFS) are the two most widely used policies for distributing tasks among available threads. Upon encountering a task creation point: (i) BFS will push the new task in a queue and continue execution of the parent task and (ii) WFS will suspend the parent task and start execution of the new task. BFS tends to be more demanding in terms of memory, as it creates all tasks before starting their execution (and thus all tasks coexist simultaneously). This is an undesirable property in general and in particular for resource-constrained embedded systems, which would make WFS a better candidate. WFS also has the nice property of following the execution path of the original sequential program, which tends to result in better data locality [5]. However, when *tied* tasks are used, BFS is the only choice in practice, as WFS leads to a complete serialization of task executions when nested parallelism is adopted. Moreover, it has been shown that the use of *untied* tasks significantly reduces the worst case response time analysis [9].

Task queue: The most widespread design solution to support the OpenMP tasking execution model is to rely on a centralized task queue. This minimizes memory footprint for runtime support metadata, which is a must in the context of embedded platforms. The basic building block of the proposed design focuses on lightweight support for *push* and *pop* operations on such a centralized queue (upon task creation and extraction, respectively), relying on fine-grained locking mechanisms. TSPs are implemented using lightweight events, which avoids the massive contention implied by active polling (idle threads on the TSP are put into sleep mode). When a task is created (i.e., pushed in the queue), the creator thread sends a signal which wakes up a single thread (selected using round-robin). After completing the task execution, the thread returns into sleeping mode. The described queue is implemented with a doubly linked list. This data structure allows to *push* and *pop* tasks from the queue and also remove a task in any position of the queue. This is key for low overhead, as tasks are not constrained to execute in-order (except when dependencies are specified), so their completion and removal from the queue is independent of their position. Note that a simple linked list does not allow this operation.

Untied tasks: The described support is sufficient to show excellent performance in the presence of simple flat parallel patterns, where all the tasks are created from within a single level (i.e., a single parent task), but lacks the capability of supporting more sophisticated forms of parallelism, like nested parallel patterns found in programs that use recursion, and for which the tasking model was originally proposed. Consequently, *untied* tasks are not supported by using this basic implementation. Due to the limitations of *tied* tasks described previously, the scheduling policy relies on BFS, and WFS is not supported. In the following, we describe how we extend this baseline implementation to fully support nested parallel patterns and *untied* tasks, while keeping the implementation lightweight and not too memory-hungry. These both are the key requirements for any implementation suitable for embedded manycore accelerators. Our main goal is to achieve a comparable efficiency in terms of task granularity (the finer the better) for which near-ideal speedups are achieved.

Figure 6.2 shows how task suspension works in most implementations supporting *tied* tasks (WFS is assumed). The thread on which the code shown in the figure is executing has an associated stack (depicted on the left). When a task directive is encountered, the thread jumps to a runtime function that manages the creation of a new task from the enclosed code region. Because WFS is considered, the thread encountering the new task executes the code encapsulated within the task region, and the parent task is suspended (as it is a tied task and so cannot migrate to a different thread). A new stack frame is activated for this task, like in every regular function call. The same thing happens at every nested task directive. When a task is completed, the stack

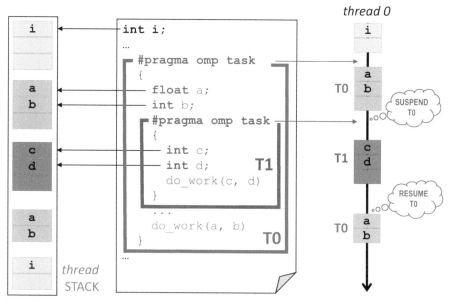

Figure 6.2 Task suspension in the baseline implementation (considering tied tasks and WFS).

pointer is reset to the top of the previous active frame. Since the semantics of *tied* task scheduling ensure that suspension/resumption can happen only on the same thread, no explicit bookkeeping to save/restore the context of a task is required.

The key extension required to support *untied* tasks is the capability of allowing to resume a suspended task on a different thread than the one that started and suspended it. To achieve this goal, we rely on lightweight *co-routines* [10]. Co-routines rely on cooperative tasks that publicly expose their code and memory state (register file, stack), so that different threads can take control of the execution after restoring the memory state. Every time that a thread suspends or resumes a suspended cooperative task, a context switch is performed. We place the required metadata to support task contexts (TCs) in the shared multi-bank memory and we use inline assembly to minimize the cost of the routines to save and restore the architectural state.

Figure 6.3 shows how task suspension works in our approach for *untied* tasks (WFS is assumed). Initially, the thread on which the code shown in the figure is executing uses its own private stack (in gray). When the outermost task region (T0) is encountered, the context of the current task is saved in the

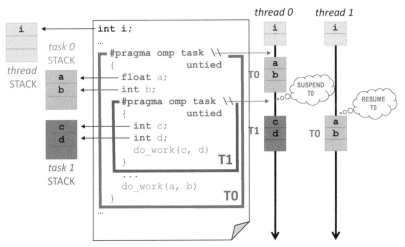

Figure 6.3 Untied task suspension with task contexts and per-task stacks.

TC (including the current SP, that is, the task pointer register), then the thread is rescheduled to execute the new task T0. The SP of the thread is updated to the stack of T0 (in blue) and the new task is started. When the creation point of the innermost task T1 is reached, an identical procedure is followed. The context of T0 is saved in its TC, which is pushed back in the queue, then thread 0 is pointed to the stack of T1 (in red). Now the suspended T0 can be pulled out of and restarted by thread 1. On top of this basic mechanism, a number of other design choices were made to minimize the cost of our runtime support, which we describe in the following.

Task hierarchy: Supporting nested tasks requires to keep in the runtime a data structure (a *tree*) that represents the hierarchy of multiple task regions. A parent task has a link to its children and vice versa, to facilitate exchange of information about execution status. For example, a parent task needs to be informed about the execution completion of its children to support the semantics of the taskwait directive. When a parent task completes its execution, its children become orphans and should not care to inform the parent. The fastest solution to handle parent task termination in terms of bookkeeping would be not to delete the descriptor, but just to maintain the task in a *zombie* status until all children have completed. This operation would require a simple update to the descriptor, which can be executed in a very short time. However, this solution brings to a memory occupation that is not acceptable for our constrained platform. Thus, we opt for a costlier removal of the descriptor from the *tree*. As a consequence, all child tasks must receive an update from the parent to avoid dangling pointers to a deallocated descriptor.

Taskwait construct: Task-level synchronization is widely used in recursive-based parallel patters. Here typically a fixed number of tasks are created at every recursion level, and their execution is synchronized with a taskwait directive. When a parent task encounters a taskwait, it should wait until all the children (first-level descendants) have completed, but typically for performance the thread hosting the parent task is allowed to switch to executing one of the children tasks. In the baseline implementation, this feature is supported by just traversing the list of children tasks in the *tree* data structure and inspecting their status to verify that it is set to WAITING. We changed this mechanism to rely on two queues per task, to directly reference children in the WAITING and RUNNING states, respectively. Upon creation, a task is inserted in the WAITING queue. Every time that a task starts to execute, the runtime moves this task from the WAITING queue to the RUNNING queue, and vice versa in case of suspension. Decoupling waiting and running tasks require a costlier bookkeeping upon task insertion and extraction, but allow faster support for taskwait as it is no longer required to search the tree for WAITING tasks. While the benefit brought by this implementation is not evident in the presence of flat parallel patterns, as the taskwait is virtually useless in this case, in recursive parallel patterns, it is extensively used and this design choice pays off.

Task dependencies: In the presence of recursive parallel patterns, it is important to distinguish between suspended tasks that could be resumed at any time and tasks that are suspended due to a scheduling constraint that needs to be unblocked. A typical example is, again, tasks suspended upon a taskwait or due to a data dependence. As already mentioned, recursive parallelism extensively relies on such a form of synchronization, thus hosting this type of suspended tasks in the same queue that also hosts ready-to-execute tasks used to lead to a situation where we would repeatedly pop from there a task just to realize that the scheduling constraint was still unsatisfied. We would then have to push back the task in the queue and retry. Checking the status of the task before extracting it does not entirely solve the problem, as it requires time-consuming search operations. To deal with this problem, we changed the implementation to avoid re-inserting in the queue suspended tasks with unresolved dependencies. Such tasks are kept floating instead, and it is up to the task that will eventually resolve the dependence to push them back into the queue. This modification requires some additional checks to deal with the above-mentioned case, but greatly improves the performance of recursive parallel programs.

Allocation of runtime metadata: To minimize the overhead for dynamic resource allocation (memory, locks, task descriptors, etc.), we have extensively used pools of pre-allocated resources. This is significantly faster than *malloc*-like primitives and does not require lock-protected operations, as we adopt thread-private resources. The downside is memory occupation. Since the targeted architecture relies on a shared cluster memory with a limited size, we have to wisely use the available space. A reasonable design solution would be to dedicate roughly 5–10% of this memory to hosting tasking support data structures. The original task descriptor has a size of 174 bytes, while the extensions that we introduced require another 98 bytes for the contexts, plus the stacks. Private thread stacks are configured to be 1 KB (a common choice for embedded systems), while task stacks are by default 1/4 of that size. Clearly, all those values are parameters in our design, and can be changed depending on specific application requirements.

Despite the increment of runtime memory requirements, the use of pre-allocated resources enables to exploit finer grained parallelism, which is paramount in current and future embedded systems. Next, we describe solutions to reduce memory pressure and runtime overhead.

Cutoff mechanisms: With 10% of the cluster's shared memory allocated to task descriptors, the runtime can host simultaneously 750 pre-allocated *tied* tasks or 400 *untied* tasks. If the queue of available task descriptors is depleted during the program execution, a mechanism (known in the literature as *cutoff* [11]) is triggered. When this condition is met, the creation of new task descriptors must be suspended to avoid that runtime resources saturate when the task production rate is greater than the execution rate. Our runtime supports two different cutoff variants: *yield* and *work-first*. In the first case, the producer task is stopped and pushed at the end of the READY queue, with the aim to re-schedule the core to executing pending tasks instead of generating new ones. Using the second variant, the producer task starts working in work-first mode by executing the new tasks in-place via a standard function call: in this case, task descriptors are not required, as the synchronization is enforced by serializing tasks on the same thread.

Cutoff mechanisms are introduced to avoid an unbounded consumption of runtime resources, but recursive applications can cause additional problems. Using *untied* tasks, task stacks typically end up to be over-sized to fit the worst case (i.e., the maximum recursion level reached in the cutoff state) to the detriment of runtime memory footprint. To avoid this case, we introduced a specific optimization for *untied* tasks using *work-first* cutoff, which forces the producer task to swap its current stack with a special one that is the only one dimensioned for worst case recursive execution.

Support for scheduling policies: The OpenMP runtime provides specific features to support the scheduling policies that have been defined in Chapter 4. Two alternative implementations are selectable for task queues: *global* and *private* queues. The global implementation defines a single task queue for the application, and it is used to support *global scheduling*. The local implementation instantiates an independent queue per thread, and it is used to support *partitioned scheduling*, in which tasks are statically allocated to threads at design time.

Adopting a *limited preemption* scheduler, each TSP in the runtime is considered as a potential preemption point. This is implemented by calling a function designed and implemented for tight integration with the RTOS. The exact behavior depends on the current scheduling policy (*global* or *partitioned*) selected for the application, which is totally transparent to the runtime.

6.3.1 Task Dependency Management

The OpenMP tasking model includes a very mature support for highly unstructured task parallelism with features to express data dependencies (on specific data elements) between tasks. To do so, OpenMP introduces the depend clause, which imposes an ordering relation between sibling tasks (tasks that are child tasks of the same task region). OpenMP defines three types of dependencies: in, out, and inout. A task with an in clause cannot start until the set of tasks with an out or an inout clause on the same data elements complete. This feature is in fact very relevant for embedded systems, often running real-time applications modeled as *direct acyclic graphs* (DAGs)[2] (see Chapter 4 for further information).

---

[2]The terms TDG and DAG are equivalent; the former is typically used when referring to runtime methodologies; the latter is used when referring to real-time analysis.

Current implementations of the OpenMP tasking model targeting the high-performance domain (e.g., libgomp, nanos++) track data dependencies among tasks by building a *task dependency graph* (TDG) at runtime. When a new task is created, its in and out dependencies are matched against those of the existing tasks. To do so, each task region maintains a *hash table* that stores the memory address of each data element contained within the out and inout clauses, and the list of tasks associated to it. The hash table is further augmented with links to those tasks depending on it, i.e., including the same data element within the in and inout clauses. In this way, when the task completes, the runtime can quickly identify its successors, which may be ready to execute.

Building the TDG at runtime requires storing the hash tables in memory until a taskwait directive is encountered. Since dependencies can be defined only between sibling tasks, when such directives are encountered, all tasks in their binding region are guaranteed to finish. Moreover, removing the information of a single task at completion would result too costly, because dependent tasks are tracked in multiple linked lists in the hash table. As a result, the memory consumption may significantly increase as the number of instantiated tasks increases.

Such a memory consumption is clearly not a problem in high-performance systems, in which large amounts of memory are available. However, this is not in general the case for parallel embedded architectures. The MPPA processor features only 2 MB of on-chip private memory per cluster. Therefore, it is paramount to devise data structures that reduce to the bare minimum the memory requirements needed to implement the TDG.

To this aim, we maintain the complete OpenMP-DAG generated by the compiler as presented in Chapter 3. *Although this idea may seem counterintuitive, the data structures needed to store a statically generated TDG are much lighter than those necessary to dynamically build the TDG.* This strategy results in a huge reduction of the memory used at runtime.

TDG Data Structure: A Sparse Matrix – A *sparse matrix* is an optimal solution to store the TDG with minimal footprint. Figure 6.4b shows the sparse matrix implementation of the DAG presented in Figure 6.4a. There, each entry contains a unique task instance identifier t_{id}, and stores in separate arrays the t_{id} and the number of tasks it depends on (labeled *Inputs* and *#in* respectively in the figure), and the t_{id} and the number of tasks depending on it (labeled *outputs* and *#out* respectively in the figure). Moreover, the sparse matrix is sorted using the t_{id}, so a dichotomic search can be applied.

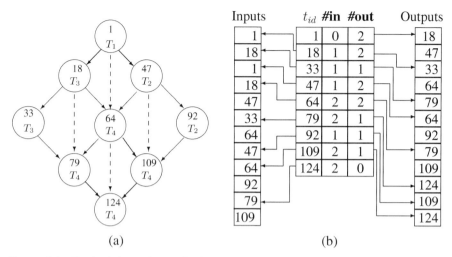

(a)   (b)

Figure 6.4   On the left (a), the DAG of an OpenMP program. On the right (b), the sparse matrix data structure implementing DAG shown on the left.

t_{id}, computed with Equation 6.1 (also presented in Chapter 3), is a key mechanism used to identify the tasks actually instantiated at runtime with those included in the DAG. Therefore, the same value of t_{id} must be generated at compile time (so each node in the DAG has a unique identifier) and at runtime (so tasks can identify its input and output data dependencies).

$$t_{id} = sid_t + T \times \sum_{i=1}^{L_t} l_i \cdot M^i \tag{6.1}$$

where sid_t is a unique task construct identifier, T is equal to the number of task, taskwait, and barrier constructs in the source code, L_t is the total number of nested loops involved in the execution of the task t, i refers to the the nesting level, l_i is the loop unique identifier at nesting level i, and M is maximum number of iterations of any considered loop.

All the information required to compute Equation 6.1 must therefore be available at compile time. sid_t is inserted by the compiler as a new parameter in the function call of the tasking runtime in charge of creating a new OpenMP task (named GOMP_task). In order to obtain the same l_i at compile-time and at runtime, the compiler introduces a *loop stack* per loop statement, and *push* and *pop* operations before the loop begins and after it ends, respectively. At every loop iteration, the top of the stack is increased by 1. The overhead associated to the stack is very little because it is inserted only in those loops where tasks are created and the overhead due to the task creation dominates. The rest of parameters, i.e., T, L_t, and M are encapsulated in the TDG data structure.

Consider task T_4, with identifier 79, in Figure 6.4a. This task instance corresponds to the computation of the matrix block $m[2, 1]$. Its identifier is computed as follows: (1) $sid_{T4} = 4$, because T_4 is the fourth task found in sequential order while traversing the source code; (2) $T = 5$ because there are four task constructs and one (implicit) barrier in the source code; (3) $L_{T_4} = 2$, the two nested loops enclosing T_4; (4) $M = 3$, the maximum number of iterations in any of the two considered loops; and (5) $l_1 = 2$ and $l_2 = 1$ are the values of the loop identifiers at the corresponding iteration. Putting all together: $T_{4_{id}} = 4 + 5(2 * 3^1 + 1 * 3^2) = 79$.

Finally, with the objective of monitoring the execution state of task instances, each entry in the sparse matrix has an associated counter (not shown in the figure) describing its state. The counter is:

- -1 if the task has not been instantiated (created) or it has finished;
- 0 if the task is ready to run; and
- > 0 if the task is waiting its input tasks to finish. The value indicates the number of tasks created and not completed it still depends on.

The runtime task scheduler works as follows:

- When a new task is created, the runtime checks the state of its *input* tasks. If all their counters are -1, the task is ready to execute; otherwise, the state of the counter of the new task is initialized with the number of input tasks with a state ≥ 0.
- When a task finishes, it decrements by 1 the counters of all its output tasks whose counter is > 0.

It is important to remark that, when the TDG contains tasks whose related if-else statement condition has not been determined at compile time and it evaluates to $false$ at runtime, the value of the counter is the same as the tasks would have already finished, i.e., -1 (see Chapter 3 for further information).

6.4 Experimental Results

In the following, we present results aimed at characterizing the overheads of the proposed OpenMP runtime design and demonstrating the reduced impact on the overall application performance, compared to different solutions.

6.4.1 Offloading Library

Synchronization on the Kalray MPPA architecture has a significant impact on the offloading cost. The preliminary implementation of the **BLOCK_OS** policy, which has the most complex semantics among all, required 75,500 cycles to initialize the runtime metadata. It is possible to halve the initialization cost by (i) replacing dynamic memory allocation of runtime data structures with a static memory mapping and (ii) distributing between the available cores the initialization of data structures.

As a further optimization, we implemented a lightweight runtime check of the presence of a pending offload request to prevent the RTOS from executing the RM when no new offload requests to process are present. This further reduced the initialization cost to 33,250 cycles. Figure 6.5 reports the offload cost on the cluster side for different synchronization policies. We report minimum and maximum observed execution cycles (blue and orange bars, respectively). The leftmost groups of bars represent the original Kalray software infrastructure, while the three rightmost groups of bars represent the three policies of our software infrastructure. The results for Kalray show a very large variance between minimum and maximum observed offload cost. Anyhow, since the analysis tools rely on worst case execution time,

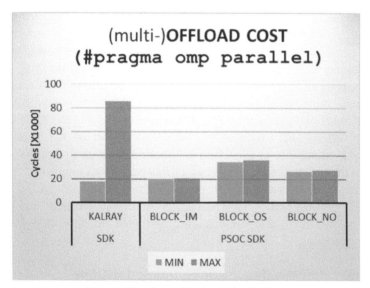

Figure 6.5 Costs of offload initialization.

to all practical purposes, we must consider the maximum time, which is around 82,000 cycles. All the three synchronization policies that we provide exhibit a very small variance, and their cost is in all cases much smaller than the worst case for the original Kalray SDK (roughly in line with the best case).

This notwithstanding, the observed costs for our runtime software are still relevant. Compared to state-of-the-art solution, we identified the main reason of this inefficiency in the management of non-coherent caches. A flush-and-invalidate operation on the data caches is performed at every synchronization point (the *unlock* primitive). This makes each access to runtime data structures very expensive in terms of execution cycles. Replacing data caches with L1 scratchpad memories and using these memories to store runtime data structures allow reducing the offload cost by 20x.

6.4.2 Tasking Runtime

As already pointed out, supporting the tasking execution model is usually subject to large overheads. While such overheads can be tolerated by large applications exploiting coarse-grained tasks, this is usually not the case for embedded applications, which rely on fine-grained workloads. To study this effect, our plots show speedup (parallel execution on 16 cluster cores versus sequential execution on a single cluster core) on the y-axis, comparing the original Kalray runtime to our runtime support for *tied* and *untied* tasks. For all the experiments except the one in Section 6.4.2.5, we use a set of microbenchmarks in which tasks only consist of ALU operations (e.g., add on local registers) and no load/store operations, which allows exploring the maximum achievable speedups. The number of ALU operations within the tasks can be controlled via a parameter, which allows studying the achievable speedup for various task granularities, which we report on the x-axis of each plot (task granularity is expressed as duration in clock cycles, roughly equivalent to the number of ALU operations that each task contains).

We consider three variants for the synthetic benchmark: LINEAR, RECURSIVE, and MIXED. These are representative of different task creation patterns found in real applications, and will be described in the following subsections.

6.4.2.1 Applications with a linear generation pattern

The LINEAR benchmark consists of $N = 512$ identical tasks, each with a workload of W ALU instructions. The main task creates all the remaining

N. . . 1 tasks from a simple loop (one task created per loop iteration) and then performs a taskwait to ensure that all tasks have completed their execution.

```
for (i=0; i<N; i++)
{
   #pragma omp task    // A task consisting
   synth (W);          // of W ALU instructions
}
#pragma omp taskwait
```

Figure 6.6 shows the results for the LINEAR benchmark. Focusing on the results for the original Kalray SDK ("noCO KALRAY" line), ideal speedups can be achieved only for tasks larger than 100 KCycles. For smaller tasks, the maximum achievable speedup is 3×. In this fine-grain task area, *our* tasks can consistently achieve a four times higher speedup. Since in the LINEAR microbenchmarks, there is no task nesting, there is no significant difference between *tied* (PSOC T) and *untied* (PSOC U) tasks. We thus explore a new configuration where tasks are recursively created to appreciate the difference.

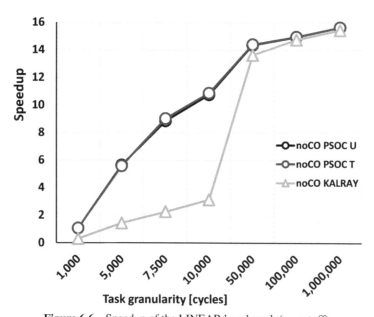

Figure 6.6 Speedup of the LINEAR benchmark (no cutoff).

6.4.2.2 Applications with a recursive generation pattern

Figure 6.7 shows the efficiency of our runtime for the recursive parallel pattern, considering *tied* and *untied* tasks. The RECURSIVE microbenchmark builds a binary tree of depth N = 9 (512 tasks) recursively. This is similar to a classical Fibonacci algorithm, where each of the two recursive calls is enclosed in a task directive. A taskwait directive is placed after the creation of the two tasks.

```
#pragma omp task    // The first task (root)
rec (0, 511);

int rec(int level, int maxlevel)
{
  if ( lev != maxlevel )
  {
    #pragma omp task          // Fist child task
    rec (level+1, maxlevel);
    #pragma omp task          // Second child task
    rec (level+1, maxlevel);
  }

  synth (W);       // W ALU instructions

  #pragma omp taskwait
}
```

The first result that we observe is that only *untied* tasks can achieve the maximum speedup. *Tied* tasks have a maximum speedup of 8. This effect is due to the behavior of taskwait in the presence of *tied* tasks. If a *tied* task is stuck on a taskwait and there are no children tasks in the WAITING state (e.g., few tasks generated at each recursion level, like in the binary tree), that task is bound to wait until the children have finished. Using a binary tree, this leads to exactly half of the threads getting stuck, which explains the maximum speedup observed in this configuration. This problem is circumvented by *untied* tasks, which can reschedule the threads hosting the stuck tasks to other ready tasks. Similar considerations to what we discussed in the previous section hold for the comparison between Kalray tasks and

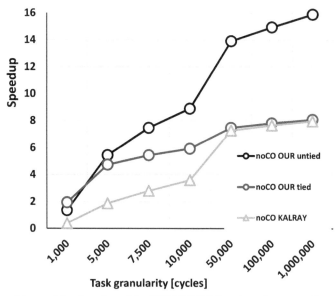

Figure 6.7 Speedup of the RECURSIVE benchmark (no cutoff).

our tied tasks (Kalray supports only *tied* tasks, so a comparison to *our untied* tasks is not directly feasible).

In general, it is possible to see that RECURSIVE implies a much higher overhead than LINEAR. This is justified by a significantly increased contention for shared data structures (queues, trees, etc.), as in this pattern multiple threads are concurrently creating tasks. Even if we have struggled to make the lock-protected operations to operate on shared data structures as short as possible, their serialization over multiple requestors is evident. As a result, it takes an order of magnitude coarser tasks (around 100 K) than in the LINEAR case to achieve nearly ideal speedups. This is a typical situation where cutoff policies can help in significantly reducing the runtime overheads. We explore the adoption of cutoff policies in Section 6.4.2.4.

6.4.2.3 Applications with mixed patterns

The advantage of using *untied* tasks is particularly evident for applications presenting a mixed structure which includes both LINEAR and RECURSIVE task creation patterns. The MIXED microbenchmark depicted in Figure 6.8

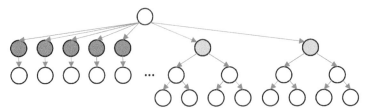

Figure 6.8 Structure of the MIXED microbenchmark.

is aimed at studying the behavior of such applications. A root task generates seven tasks in a LINEAR manner, each one spawning a single child with a long execution time and then performing a taskwait, plus another two tasks from within RECURSIVE binary trees of depth 5.

Figure 6.9 shows the results for this benchmark. Using *tied* tasks, 14 threads are allocated to execute the linear part of the application, seven of which are blocked by the taskwait directive. The ideal speedup of the application is 2, which our *tied* tasks reach for granularities of around 10 Kcycles.

Using *untied* tasks, only seven threads are allocated to the LINEAR part, which brings the ideal speedup to 9×. The maximum speedup achieved by our *untied* tasks is 8, due to a limitation of the tracing (performance

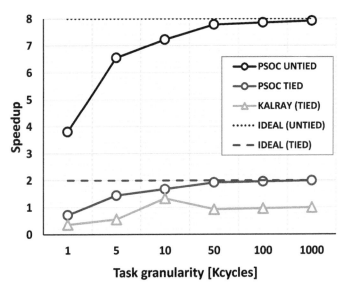

Figure 6.9 Speedup of the MIXED benchmark.

monitoring) of the Kalray platform. The root task of the hierarchy is the one performing time measurement and we were forced to declare this as a *tied* task to gather coherent clock values (allowing this task to migrate to other cores results in incoherent measurement). This limits the maximum achievable speedup to $8\times$, which our *untied* tasks achieve for granularities above 10 Kcycles.

Overall, *untied* tasks enable four times faster execution than *tied* tasks for application featuring mixed task creation patterns. Note that this result holds for any runtime implementation. Our solution makes this result visible for smaller tasks compared to other OpenMP tasking implementations. The Kalray implementation never enables any speedup in the considered range of task granularities (up to one million cycles) for this experiment.

6.4.2.4 Impact of cutoff on LINEAR and RECURSIVE applications

We repeated the experiments with LINEAR and RECURSIVE microbenchmarks considering a higher number of tasks (2,048). This configuration saturates the runtime data structures and activates *cutoff* mode. Figures 6.10 and 6.11 show the results for this experiment.

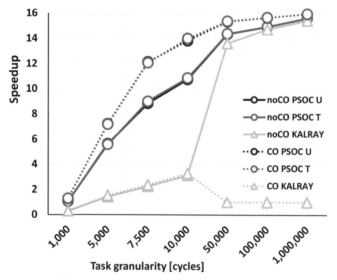

Figure 6.10   Speedup of the LINEAR benchmark (with cutoff).

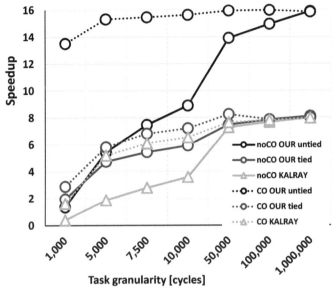

Figure 6.11   Speedup of the RECURSIVE benchmark (with cutoff).

Focusing on the LINEAR pattern, the adoption of cutoff greatly mitigates overhead effects, and we can achieve nearly ideal speedups for an order of magnitude smaller tasks compared to Kalray tasks. It also has to be noted that cutoff mode is not properly supported for LINEAR patterns in the original Kalray runtime. Enabling cutoff mode in this configuration simply seems to disable parallelism completely. Focusing on the RECURSIVE pattern, the use of cutoff policies proves extremely beneficial, with nearly ideal speedups for very fine-grained tasks (in the order of thousand cycles).

6.4.2.5 Real applications

To assess the performance of our tasking runtime on real applications, we execute the benchmarks from the Barcelona OpenMP Task Suite (BOTS) [12], which includes a wide set of real-life applications parallelized with OpenMP tasks.

Figure 6.12 shows the speedup of applications for different configurations, comparing the Kalray SDK (KALRAY) with different configurations of our runtime, using *tied* tasks (PSOC tied), *untied* tasks (PSOC untied), and *untied* tasks with cutoff (PSOC untied CO2).

On average, programs executing on top of our runtime show a speedup of $12\times$, compared to only $8\times$ for the original Kalray SDK. The benefits of cutoff

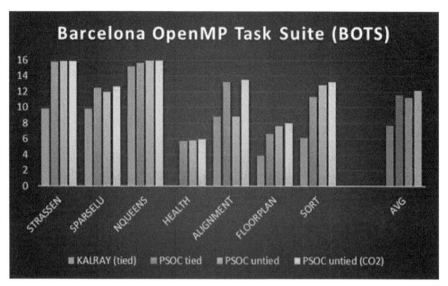

Figure 6.12 Speedups for the BOTS benchmarks.

here are minimal, since the bottleneck is limited parallelism in the application rather than runtime overhead. The marginal improvements enabled by cutoff, where present, are usually due to better memory usage (tasks in cutoff use less memory for the runtime, which is used for application data instead).

6.4.3 Evaluation of the Task Dependency Mechanism

This section evaluates the use of a sparse matrix to implement the TDG upon which the task dependency mechanism is built as presented in Section 6.3.1.

Concretely, we implement our task dependency mechanism on top of the GNU libgomp library included in GCC version 4.7.2, which supports tasks but not dependencies, and compare it with the libgomp library included in GCC 4.9.2, which implements a dependency checker based on a hash table structure.

The reason to implement our mechanism on a library not supporting dependencies is that both implementations differ only in the dependency checker, and so being easier to incorporate a new one, rather than replacing it. Moreover, to ensure that results are not affected by the version of the library, we executed the applications considered in this section without dependence clauses. Despite the incorrect result, the numbers revealed that both libraries have the exact same memory usage and performance, demonstrating that the memory increment is exclusively caused by using different dependency checkers.

Moreover, we consider two applications, one from the HPC domain, i.e., a *cholesky factorization* [13] used for efficient linear equation solvers and Monte Carlo simulations, and one from the embedded domain, i.e., an application resembling the *3D path planning* [14] (r3DPP) used for airborne collision avoidance.

For comparison purposes, the applications have been parallelized with task dependencies, i.e., using the depend clause, and without dependencies, i.e., using only task and taskwait directives.

6.4.3.1 Performance speedup and memory usage

Figures 6.13 and 6.14 show the performance speedup and the runtime memory usage (in KB) of the Cholesky and r3DPP, when varying the number of instantiated tasks, ranging from 1 to 5984 and 4096, respectively, and considering the three libgomp runtimes implementing a dependency checker based on a hash table, on a sparse matrix, and one with not dependency checker (labeled *omp4*, *omp 3.1*, and *lightweight omp4*, respectively).

The performance has been computed with the average of 100 executions. Similarly, Figures 6.14a,b show the heap memory usage (in KB) of the three OpenMP runtimes when executing Cholesky and r3DPP respectively and varying the number of instantiated tasks as well. The memory usage has been extracted using *Valgrind Massif* [15] tool, which allows profiling the heap memory consumed by the runtime in which the TDG structure is maintained.

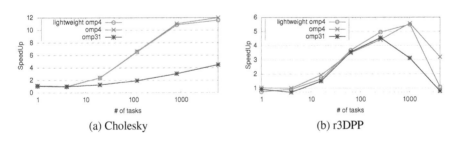

(a) Cholesky   (b) r3DPP

Figure 6.13 Performance speedup of the Cholesky (a) and r3DPP (b) running with *lightweight omp4, omp4*, and *omp 3.1*, and varying the number of tasks.

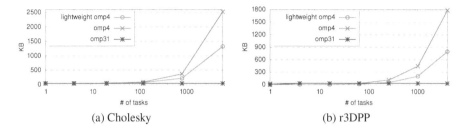

(a) Cholesky   (b) r3DPP

Figure 6.14   Memory usage (in KB) of the Cholesky (a) r3DPP (b) running with *lightweight omp4, omp4*, and *omp 3.1*, and varying the number of tasks.

For these experiments, we consider an Intel Xeon CPU E5-2670 processors, featuring eight cores each, with 20 MB L3. The reason is that it incorporates the libgomp library included in GCC 4.9.2 supporting dependency checker based on a hash table.

We observe that both performance and memory usage depend on the number of instantiated tasks: the higher the number of instances, the better the performance, as the chances of parallelism increase. When the number of tasks is too high, however, the overhead introduced by the runtime and the small workload of each task slows down the performance.

As shown in Figure 6.13, our *lightweight omp4* obtains the same performance speedups as the *omp4* implementation for the two applications, and outperforms *omp 3.1*. However, when observing the memory usage in Figure 6.14, it rapidly increases for *omp4*, requiring much more memory than the runtime based on the sparse matrix, i.e., the *lightweight omp4*.

It is also interesting to observe the parallelization opportunities brought by the depend clause, which makes the performance of Cholesky (Figure 6.13a) to increase significantly compared to not using them, with a speedup increment from 4x to 12x when instantiating 5,984 tasks. At this point, *omp4* consumes 2.5 MB while our *lightweight omp4* requires less than 1.3 MB. The memory consumed by *omp3.1* is less than 100 KB (Figure 6.14a). In fact, the *omp3.1* memory consumption is similar for all the applications because no structure for dependencies management is needed.

For the r3DPP, the depend clause achieves a performance speedup of 5.2x and 5.8x with *omp4* and *lightweight omp4*, respectively, when instantiating 1,024 tasks (Figure 6.13b). At this point, *omp4* consumes 400 KB in front of the 200 KB consumed by *lightweight omp4* (Figure 6.14b). Not considering dependencies, i.e., *omp31*, achieves a maximum performance of 4.5x when 256 tasks are instantiated (Figure 6.13b). When the number of task instances

Table 6.1 Memory usage of the sparse matrix (in KB), varying the number of tasks instantiated

Cholesky	Tasks	4	20	120	816	5984
	KB	0.11	0.59	3.80	27.09	204.19

r3DPP	Tasks	16	64	256	1024	4096
	KB	00.47	1.94	7.88	31.75	127.5

increases to 4096, all runtimes suffer a significant performance degradation because the number of instantiated tasks is too high compared to the workload computed by each task.

Table 6.1 shows the size of the sparse matrix data structure implementing the esTDG of each application when varying the number of instantiated tasks (the memory consumption reported in Figures 6.14a,b already includes it).

6.4.3.2 The task dependency mechanism on the MPPA

To evaluate the benefit of the task dependency mechanism on a memory constrained manycore architecture, we evaluated it on the MPPA processor. Figure 6.15 shows the performance speedup of Cholesky (a) and r3DPP (b) executed in one MPPA cluster, considering the *lightweight omp4* and *omp31* runtimes and varying the number of tasks. Note that *omp4* runtime experiments are not provided because MPPA does not support it. Memory consumption is the same as the one shown in Figure. 6.14 r3DPP increases the performance speedup from 9x to 12x when using our *lightweight omp4* rather than *omp3.1* and only consuming 200 KB. Cholesky presents a significant speedup increment when instantiating 816 tasks, i.e., from 2.5x to 9x, consuming only 220 KB.

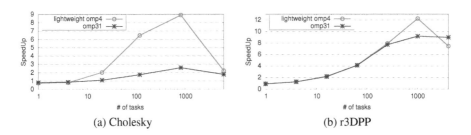

(a) Cholesky   (b) r3DPP

Figure 6.15 Performance speedup of the Cholesky (a) and r3DPP (b) running on the MPPA with *lightweight omp4, omp4,* and *omp 3.1*, and varying the number of tasks.

6.5 Summary

This chapter has illustrated the design of the OpenMP runtime for a heterogeneous platform including a *host* processor and an embedded manycore accelerator. The complete software stack is composed of an offloading library and a tasking runtime library, which have been described in detail. The OpenMP runtime provides specific features to support the scheduling policies that have been defined in Chapter 4, and it also implements the TDG required to support the task dependency mechanism as presented in Section 6.3.1. The chapter has discussed how to enable maximum exploitation of the available hardware parallelism via the *untied* task model, highlighting the key design choices to achieve low overhead. Experimental results show that this enables up to four times faster execution than *tied* tasks, which improves on average by 60% over the native Kalray SDK.

References

[1] Marongiu, A., Capotondi, A., Tagliavini, G., and Benini, L., "Simplifying Many-Core-Based Heterogeneous SoC Programming With Offload Directives." In *IEEE Transactions on Industrial Informatics*, vol. 11, pp. 957–967, 2015.

[2] Mitra, G., Stotzer, E., Jayaraj, A., and Rendell, A. P., "Implementation and optimization of the OpenMP accelerator model for the TI Keystone II architecture." In *International Workshop on OpenMP*, Springer, pp. 202–214, 2014.

[3] Rosenstiel, W., and Thiele, L. (editors), *Design, Automation and Test in Europe Conference and Exhibition, DATE 2012, Dresden, Germany.* IEEE, 2012.

[4] Podobas, A., Brorsson, M., and Faxén, K.-F., A comparative performance study of common and popular task-centric programming frameworks. *Concurr. Comput. Pract. Exp.* 27, 1–28, 2015.

[5] Duran, A., Teruel, X., Ferrer, R., Martorell, X., and Ayguade, E., "Barcelona OpenMP Tasks Suite: A Set of Benchmarks Targeting the Exploitation of Task Parallelism in OpenMP." In *2009 International Conference on Parallel Processing*, pp. 124–131. IEEE, 2009.

[6] Burgio, P., Tagliavini, G., Marongiu, A., and Benini, L., "Enabling fine-grained OpenMP tasking on tightly-coupled shared memory clusters." In *Proceedings of the Conference on Design, Automation and Test in Europe*, DATE '13, pp. 1504–1509. EDA Consortium, 2013.

[7] Rochange, C., Bonenfant, A., Sainrat, P., Gerdes, M., Lobo, J., et al., "WCET analysis of a parallel 3D multigrid solver executed on the MERASA multi-core." In *WCET*, 2010.

[8] Kumar, S., Hughes, C. J., and Nguyen, A, "Carbon: Architectural Support for Fine-grained Parallelism on Chip Multiprocessors." In *Proceedings of the 34th Annual International Symposium on Computer Architecture*, ISCA '07, pp. 162–173. ACM, 2007.

[9] Serrano, M. A., Melani, A., Vargas, R., Marongiu, A., Bertogna, M., and Quiñones, E., "Timing Characterization of OpenMP4 Tasking Model." In *Proceedings of the 2015 International Conference on Compilers, Architecture and Synthesis for Embedded Systems*, CASES '15, pp. 157–166. IEEE Press, 2015.

[10] Marlin, C. D., *Coroutines: a programming methodology, a language design and an implementation*. Number 95 in Lecture Notes in Computer Science. Springer Science and Business Media, 1980.

[11] Duran, A., Corbalán, J., and Ayguadé, E., "Evaluation of OpenMP task scheduling strategies." In *International Workshop on OpenMP*, pp. 100–110. Springer, 2008.

[12] Duran, A., Corbalan, J., and Ayguade, E., "An adaptive cut-off for task parallelism." In *2008 SC – International Conference for High Performance Computing, Networking, Storage and Analysis*, pp. 1–11. IEEE, 2008.

[13] Bascelija, N., Sequential and Parallel Algorithms for Cholesky Factorization of Sparse Matrices. *WSEAS: Mathematic. Appl. Sci. Mech.* 2013.

[14] Cesarini, D., Marongiu, A., and Benini, L., "An optimized task-based runtime system for resource-constrained parallel accelerators." In *2016 Design, Automation and Test in Europe Conference and Exhibition, DATE 2016*, Dresden, Germany, pp. 1261–1266, 2016.

[15] Nethercote, N., et. al., "Building Workload Characterization Tools with Valgrind." In *IISWC*, 2006.

7

Embedded Operating Systems

**Claudio Scordino[1], Errico Guidieri[1], Bruno Morelli[1],
Andrea Marongiu[2,3], Giuseppe Tagliavini[3] and Paolo Gai[1]**

[1]Evidence SRL, Italy
[2]Swiss Federal Institute of Technology in Zurich (ETHZ), Switzerland
[3]University of Bologna, Italy

In this chapter, we will provide a description of existing open-source operating systems (OSs) which have been analyzed with the objective of providing a porting for the reference architecture described in Chapter 2. Among the various possibilities, the ERIKA Enterprise RTOS (Real-Time Operating System) and Linux with preemption patches have been selected. A description of the porting effort on the reference architecture has also been provided.

7.1 Introduction

In the past, OSs for high-performance computing (HPC) were based on custom-tailored solutions to fully exploit all performance opportunities of supercomputers. Nowadays, instead, HPC systems are being moved away from in-house OSs to more generic OS solutions like Linux. Such a trend can be observed in the TOP500 list [1] that includes the 500 most powerful supercomputers in the world, in which Linux dominates the competition. In fact, in around 20 years, Linux has been capable of conquering all the TOP500 list from scratch (for the first time in November 2017).

Each manufacturer, however, still implements specific changes to the Linux OS to better exploit specific computer hardware features. This is especially true in the case of computing nodes in which lightweight kernels are used to speed up the computation.

System Count

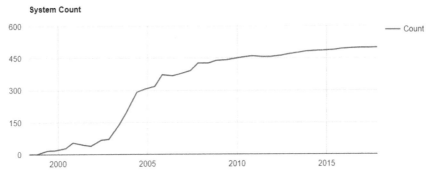

Figure 7.1   Number of Linux-based supercomputers in the TOP500 list.

Linux is a full-featured OS, originally designed to be used in server or desktop environments. Since then, Linux has evolved and grown to be used in almost all computer areas – among others, embedded systems and parallel clusters. Linux currently supports almost every hardware processor, including x86, x86-64, ARM, PowerPC, MIPS, SuperH, IBM S/390, Motorola 68000, SPARC, etc. The programmability and the portability of code across different systems are ensured by the well-known "Portable Operating System Interface" (POSIX) API. This is an IEEE standard defining the basic environment and set of functions offered by the OS to the application programs.

Hence, the main reason for this success and popularity in the HPC sector is its excellent performance and its extreme scalability, due to very carefully designed data structures like Linux Read-Copy Update (RCU) [2]. This scalability, together with the high modularity, enables excellent performance on both a powerful parallel cluster made by thousands of cores and a small embedded microcontroller, as will be shown in the next sections.

Therefore, when designing the support for our predictable parallel programming framework, we started selecting Linux as the basic block for executing the target parallel applications. On the other hand, Linux alone is not sufficient for implementing the needed runtime support on our reference architecture: a solution needed to be found for the compute cores, where a tiny RTOS is needed in order to provide an efficient scheduling platform to support the parallel runtime described in Chapter 6.

This chapter in particular describes in detail how the scheduling techniques designed in Chapter 4 have been implemented on the reference architecture. The chapter includes notes about the selection of the tiny RTOS for the compute cores, with a description of the RTOS, as well as the solutions implemented to support Linux on the I/O cores with real-time performance.

This chapter is structured as follows. Section 7.2 describes the state of the art of the real-time support for the Linux OS and as well for small RTOSes. Section 7.3 describes the requirements that influenced the choice of the RTOS, which is described in detail in Section 7.4. Section 7.5 provides some insights about the OS support for the host processor and for the many-core processor. Finally, Section 7.6 summarizes the chapter.

7.2 State of The Art

7.2.1 Real-time Support in Linux

As noted in the Section "Introduction," in the last years, there has been a considerable interest in using Linux for both HPC and real-time control systems, from academic institutions, independent developers, and industries. There are several reasons for this rising interest.

First of all, Linux is an Open Source project, meaning that the source code of the OS is freely available to everybody, and can be customized according to user needs, provided that the modified version is still licensed under the GNU General Public License (GPL) [3]. This license allows anybody to redistribute, and even sell, a product as long as the recipient is able to exercise the same rights (access to the source-code included). This way, a user (for example, a company) is not tied to the OS provider anymore, and is free to modify the OS at will. The Open Source license helped the growth of a large community of researchers and developers who added new features to the kernel and ported Linux to new architectures. Nowadays, there is a huge number of programs, libraries, and tools available as Open Source code that can be used to build a customized version of the OS.

Moreover, Linux has the simple and elegant design of the UNIX OSs, which guarantees meeting the typical reliability and security requirements of real-time systems.

Finally, the huge community of engineers and developers working on Linux makes finding expert programmers very easy.

Unfortunately, the standard mainline kernel (as provided by Linus Torvalds) is not adequate to be used as RTOS. Linux has been designed to be a general-purpose operating system (GPOS), and thus not much attention has been given to the problem of reducing the latency of critical operations. Instead, the main design goal of the Linux kernel has been (and still remains) to optimize the average throughput (i.e., the amount of "useful work" done by the system in the unit of time). For this reason, a Linux program may suffer a high latency in response to critical events. To overcome these problems, many approaches have been proposed in the last years to modify Linux in order to make it more "real-time." These approaches can be grouped in the following classes [4]:

1. Hard real-time scheduling through a Hardware Abstraction Layer (HAL);
2. Latency reduction through better preemption mechanisms and interrupt handling;
3. Proper real-time scheduling policies.

The following subsections describe each approach in detail.

7.2.1.1 Hard real-time support

This approach consists in creating a layer of virtual hardware between the standard Linux kernel and the real hardware. This layer is called Real-Time Hardware Abstraction Layer (RTHAL). It abstracts the hardware timers and interrupts and adds a separate subsystem to run the real-time tasks. The Linux kernel and all the normal Linux processes are then managed by the abstraction layer as the lowest priority tasks — i.e., the Linux kernel only executes when there are no real-time tasks to run.

The first project implementing this approach was RTLinux [5]. The project started at Finite State Machine Labs (FSMLabs) in 1995. Then, it was released in two different versions: an Open Source version (under GPL license) and a more featured commercial version. An operation of patenting issued in US in 1999, however, generated a massive transition of developers towards the parallel project RTAI. Then, the commercial version was bought by WindRiver. Nowadays, both versions are not maintained anymore [5].

RTAI [6] (which stands for "Real-Time Application Interface") is a project started as a variant of RTLinux in 1997 at Dipartimento di Ingegneria Aerospaziale of Politecnico di Milano (DIAPM), Italy. The project is under LGPL license, and it was supported by a large community of developers, based on the Open Source model. Although the project initially started from the original RTLinux code, it has been completely rewritten over time. In particular, the RTAI community has developed the Adaptive Domain Environment for Operating Systems (ADEOS) nanokernel as an alternative for RTAI's core, to get rid of the old kernel patch and exploit a more structured and flexible way to add a real-time environment to Linux [4]. The project mainly targets the x86 architecture and is currently maintained (even if less popular than it used to be in the past).

Xenomai [7] was born in 2001 as an evolution of Fusion, a project to run RTAI tasks in the user space. With Xenomai, a real-time task can execute in user space or in kernel space. Normally, it starts in kernel space (i.e., "primary domain"), where it has real time performance. When the real-time task invokes a function belonging to the Linux standard API or libraries, it is automatically migrated to the user-level (i.e., "secondary domain"), under the control of the Linux scheduler. In this secondary domain, it keeps a high priority, being scheduled with the SCHED FIFO or SCHED RR Linux policies. However, it can experience some delay and latency, due to the fact that it is scheduled by Linux. After the function call has been completed, the task can go back to the primary mode by explicitly calling a function. In this way, at the cost of some limited unpredictability, the real-time programmer can use the full power of Linux also for real-time applications.

Among the various projects implementing the hardware abstraction approach, Xenomai is the one which supports the highest number of embedded architectures. It supports ARM, Blackfin, NiosII, PPC and, of course, x86. Xenomai also offers a set of skins implementing the various APIs of popular RTOS such as Windriver VxWorks [8], as well as the POSIX API [9]. In version 3 of Xenomai, the project aims at working on top of both a native Linux kernel and a kernel with PREEMPT_RT [10], by providing a set of user-space libraries enabling seamless porting of applications among the various OS versions.

It is important to highlight the advantages of the approach of hardware abstraction. First of all, the latency reduction is really effective [4]. This allows the implementation of very fast control loops for applications like vibrational control. Moreover, it is possible to use a full-featured OS like Linux for both the real-time and the non-real-time activities (e.g., HMI, logging, monitoring, communications, etc.). Finally, the possibility of developing and then executing the code on the same hardware platform, considerably simplifies the complexity of the development environment.

Typical drawbacks of this approach – which depend on the particular implementation – are:

- Real-time tasks must be implemented using specific APIs, and they cannot access typical Linux services without losing their real-time guarantees.
- The implementation is very hardware-dependent, and may not be available for a specific architecture.
- The real-time tasks are typically executed as modules dynamically loaded into the kernel. Thus, there is no memory protection and a buggy real-time task may crash the whole system.

For these reasons, this approach is usually followed only to build hard real-time systems with very tight requirements.

7.2.1.2 Latency reduction

"Latency" can be defined as the time between the occurrence of an event and the beginning of the action to respond to the event [4]. In the case of an OS, it is often defined as the time between the interrupt signal arriving to the processor (signaling the occurrence of an external event like data from a sensor) and the time when the handling routine starts execution (e.g., the real-time task that responds to the event). Since in the development of critical real-time control systems, it is necessary to account for the worst-case scenario, a particularly important measure is the maximum latency value.

The two main sources of latency in general-purpose OSs are task latency and timer resolution:

1. Task latency is experienced by a process when it cannot preempt a lower priority process because this is executing in kernel context (i.e., the kernel is executing on behalf of the process). Typically, monolithic OSs do not allow more than one stream of execution in kernel context, so that the high-priority task cannot execute until the kernel code either returns to user-space or explicitly blocks. As we will explain in the following paragraphs, Linux has been capable of mixing the advantages of a traditional monolithic design with the performance of concurrent streams of execution within the kernel.
2. The timer resolution depends on the frequency at which the electronics issues the timing interrupts (also called "tick"). This hardware timer is programmed by the OS to issue interrupts at a pre-programmed period of time. The periodic tick rate directly affects the granularity of all timing activities. The Linux kernel has recently switched towards a dynamic tick timer, where the timer does not issue interrupts at a periodic rate. This feature allows the reduction of energy consumption whenever the system is idle.

In the course of the years, several strategies have been designed and implemented by kernel developers to reduce these values. Among the mechanisms already integrated in the official Linux kernel, we can find:

- Robert Love's Preemptible Kernel patch to make the Linux kernel preemptible just like user-space. This means that several flows of kernel execution can be run simultaneously. Urgent events can be served regardless of the fact that the system is running in the kernel context.

Hence, it becomes possible to preempt a process at any point, as long as the kernel is in a consistent state. With this patch the Linux kernel has become a fully preemptive kernel, unlike most existing OSs (UNIX variants included). This feature was introduced in the 2.6 kernel series (December 2003).

- High Resolution Timers (HRT) is a mechanism to issue timer interrupts aperiodically – i.e., the system timer is programmed to generate the interrupt after an interval of time that is not constant, but depends on the next event scheduled by the OS. Often, these implementations also exploit processor-specific hardware (like the APIC on modern x86 processors) to obtain a better timing resolution. This feature was introduced in the 2.6.16 kernel release (March 2006).

- Priority inheritance for user-level mutex, available since release 2.6.18 (September 2006). Priority inheritance support is useful to guarantee bounded blocking times in case more than one thread needs to concurrently access the same resource. The main idea is that blocking threads inherit the priority of the blocked threads, thus giving them additional importance in order to finish their job early.

- Threaded interrupts by converting interrupt handlers into preemptible kernel threads, available since release 2.6.30 (June 2009). To better understand the effect of this patch, we have to consider that the typical way interrupts are managed in Linux is to manage the effect of the interruption immediately inside the so-called interrupt handler. In this way, peripherals are handled immediately, typically providing a better throughput (because thread waiting for asynchronous events are put earlier in the ready queue). On the other hand, a real-time system may have a few "important" IRQs that need immediate service, while the others, linked to lower priority activities (e.g., network, disk I/O), can experiences higher delays. Therefore, having all interrupt services immediately may provide unwanted jitter in the response times, as low-priority IRQ handlers may interrupt high-priority tasks. The threaded interrupt patch solves this problem by transforming all IRQ handlers into kernel threads. As a result, the IRQ handlers (and their impact on the response time) are minimized. Moreover, users can play with real-time priorities to eventually raise the priorities of the important interrupts, therefore providing stronger real-time guarantees.

PREEMPT_RT [10] is an on-going project supported by the Linux Foundation [11] to bring real-time performance to a further level of sophistication, by introducing preemptible ("sleeping") spinlocks and RT mutexes implementing Priority Inheritance to avoid priority inversion.

It is worth specifying that the purpose of the PREEMPT_RT patch is not to improve the throughput or the overall performance. The patch aims at reducing the maximum latency experienced by an application to make the system more predictable and deterministic. The average latency, however, is often increased.

7.2.1.3 Real-time CPU scheduling

Linux systems traditionally offered only two kind of scheduling policies:

1. SCHED_OTHER: Best-effort round-robin scheduler;
2. SCHED_FIFO/SCHED_RR: Fixed-priority POSIX scheduling.

During the last decade, due to the increasing need of a proper real-time scheduler, a number of projects have been proposed to add more sophisticated real-time scheduling (e.g., SCHED_SOFTRR [12], SCHED_ISO [13], etc.). However, they remained as separate projects and have never been integrated in the mainline kernel.

During the last years, the real-time scheduler SCHED_DEADLINE [14, 15] originally proposed and developed by Evidence Srl in the context of the EU project ACTORS [16], has been integrated in the Linux kernel. It is available since the stable release 3.14 (March 2014). It consists of a platform-independent real-time CPU scheduler based on the Earliest Deadline Scheduler (EDF) algorithm [17], and it offers temporal isolation between the running tasks. This means that the temporal behavior of each task (i.e., its ability to meet its deadlines) is not affected by the behavior of the other tasks running in the system. Even if a task misbehaves, it is not able to exploit larger execution times than the amount it has been allocated. The scheduler only enforces temporal isolation on the CPU, and it does not yet take into account shared hardware resources that could affect the timing behavior.

A recent collaboration between Scuola Superiore Sant'Anna, ARM Ltd. and Evidence Srl, has aimed at overcoming the non-work-conserving nature of SCHED_DEADLINE while keeping the real-time predictability. This joint effort that replaced the previous CBS algorithm with GRUB has been merged since kernel release 4.13.

7.2.2 Survey of Existing Embedded RTOSs

The market of embedded RTOSs has been exploited in the past decades by several companies that have been able to build solid businesses. These companies started several years ago, when the competition from free OSs was non-existent or very low. Thus, they had enough time to create a strong business built on top of popular and reliable products. Nowadays, the market is full of commercial solutions, which differentiate in the application domain (e.g., automotive, avionics, railway, etc.) and in the licensing model. Most popular commercial RTOSs are: Windriver VxWorks [8], Green Hills Integrity [18], QNX [19], SYSGO PikeOS [20], Mentor Graphics Nucleus RTOS [21], LynuxWorks LynxOS [22], and Micrium μc/OS-III [23]. However, there are some other interesting commercial products like Segger EmbOS [24], ENEA OSE [25], and Arccore Arctic core [26].

On the other hand, valid Open-Source alternatives exist. The development of a completely free software tool chain being our target, the focus of this subsection will be more on the free RTOSs available publicly. Some free RTOSs, in fact, have now reached a level of maturity in terms of reliability and popularity that can compete with commercial solutions. The Open-Source licenses allow the modification of the source code and porting the RTOS on the newest many-core architectures.

This section provides an overview of the free RTOSs available. For each RTOS, the list of supported architectures, the level of maturity and the kind of real-time support are briefly provided. Other information about existing RTOSs can be found in [27].

FreeRTOS

FreeRTOS [28] is a small RTOS written in C. It provides threads, tasks, mutexes, semaphores and software timers. A tick-less mode is provided for low-power applications.

It supports several architectures, including ARM (ARM7/9, Cortex-A/M), Altera Nios2, Atmel AVR and AVR32, Cortus, Cypress PSoC, Freescale Coldfire and Kinetis, Infineon TriCore, Intel x86, Microchip dsPIC and PIC18/24/32 and dsPIC, Renesas H8/S, SuperH, Fujitsu, Xilinx Microblaze, etc.

It does not implement very advanced scheduling algorithms, but it offers a classical preemptive or cooperative fixed-priority round-robin with priority inheritance mutexes.

The RTOS is Open Source, and was initially distributed under a license similar to GPL with linking exception [29]. Recently the FreeRTOS kernel has been relicensed under the MIT license thanks to the collaboration with Amazon AWS. A couple of commercial versions called SafeRTOS and OpenRTOS are available as well. The typical footprint is between 5 KB and 10 KB.

Contiki

Contiki [30] is an Open-Source OS for networked, memory-constrained systems with a particular focus on low-power Internet of things devices. It supports about a dozen microcontrollers, even if the ARM architecture is not included. The Open-Source license is BSD, which allows the usage of the OS in commercial devices without releasing proprietary code.

Although several resources include Contiki in the list of free RTOSs, Contiki is not a proper RTOS. The implementation is based on the concept of protothreads, which are non-preemptible stack-less threads [31]. Context switch can only take place on blocking operations, and does not preserve the content of the stack (i.e., global variables must be used to maintain variables across context switches).

Stack sharing is a useful feature, but the lack of preemptive support and advanced scheduling mechanisms made this OS not suitable to meet the needs of the parallel programming software framework we want to implement.

Marte OS

Marte OS [32] is a hard RTOS that follows the Minimal Real-Time POSIX.13 subset. It has been developed by the University of Cantabria. Although it is claimed to be designed for the embedded domain, the only supported platform is the x86 architecture. The development is discontinued, and the latest contributions date back to June 2011.

Ecos and EcosPro

Ecos [33] is an Open-Source RTOS for applications which need only one process with multiple threads. The source code is under GNU GPL with linking exception.

The current version is 3.0 and it runs on a wide variety of hardware architectures, including ARM, CalmRISC, Motorola 68000/Coldfire, fr30, FR-V, Hitachi H8, IA32, MIPS, MN10300, OpenRISC, PowerPC, SPARC, SuperH, and V8xx.

The footprint is in the order of tens of KB, which does not make it suitable for processing units with extremely low memory. The kernel is currently developed in a closed-source fork named eCosPro.

FreeOSEK

FreeOSEK [34] is a minimal RTOS implementing the OSEK/VDX automotive standard, like Erika Enterprise. The Open-Source license (GNU GPLv3 with linking exception) is similar to the one of Erika Enterprise too. However, it only supports the ARM7 architecture, the development community is small, and the project does not appear to be actively maintained.

QP

Quantum platform (QP) [35] is a family of lightweight, open source software frameworks developed by company Quantum Leaps. These frameworks allow building modular real-time embedded applications as systems of cooperating, event-driven active objects (actors). In particular, QK (Quantum Kernel) is a tiny preemptive non-blocking run-to-completion kernel designed specifically for executing state machines in a run-to-completion (RTC) fashion.

Quantum platform supports several microcontrollers, including ARM Cortex-M, ARM 7/9/Cortex-M, Atmel AVR Mega and AVR32, Texas Instruments MSP430/TMS320C28x/TMS320C55x, Renesas Rx600/R8C/H8, Freescale Coldfire/68HC08, Altera Nios II, Microchip PIC24/dsPIC, and Cypress PSoC1.

The software is released in dual licensing: an Open-Source and a commercial license. The Open-Source license is GNU GPL v3, which requires the release of the source code to any end user. Unfortunately, the Open-Source license chosen is not suitable for consumer electronics, where the companies want to keep the intellectual property of their application software.

Trampoline

Trampoline [36] is an RTOS which aims at OSEK/VDX automotive certification. However, unlike ERIKA Enterprise, it has not yet been certified.

Only the following architectures are supported: Cortex M, Cortex A7 (alone or with the Hypervisor XVisor), RISC-V, PowerPC 32 bits, AVR, ARM 32 bit.

The Open-Source license at the time the evaluation was made was LGPL v2.1. This license is not very suitable for consumer electronics because it implies that any receiver of the binary (e.g., final user buying a product) must be given access to the low-level and the possibility of relinking the application towards a newer version of the RTOS. The license was changed afterwards to GPL v2 in September 2015.

RTEMS

RTEMS [37] is a fully-featured Open-Source RTOS supporting several application programming interfaces (APIs) such as POSIX and BSD sockets. It is used in several application domains (e.g., avionics, medical, networking) and supports a wide range of architectures including ARM, PowerPC, Intel, Blackfin, MIPS, and Microblaze. It implements a single process, multi-threaded environment. The Open-Source license is similar (but not equal) to the more popular GPL with Linking Exception [29].

The footprint is not extremely small, and for the smallest applications, ranges from 64 to 128 K on nearly all CPU families [38]. For this reason, another project called TinyRTEMS [39] has been created to reduce the footprint of RTEMS. However, its Open-Source license is GPLv2, which is not suitable for development in industrial contexts.

TinyOS

TinyOS [40] is an Open-Source OS specifically designed for low-power wireless devices (e.g., sensor networks) and mainly used in research institutions. It has been designed for very resource-constrained devices, such as microcontrollers with a few KB of RAM and a few tens of KB of code space. It's also been designed for devices that need to be very low power.

TinyOS programs are built out of software components, some of which present hardware abstractions. Components are connected to each other using interfaces. TinyOS provides interfaces and components for common abstractions such as packet communication, routing, sensing, actuation, and storage.

TinyOS cannot be considered a proper real-time OS, since it implements a non-preemptive thread model.

The OS is licensed under BSD license which, like GPL with linking exception, does not require redistribution of the source code of the application.

TinyOS supports Texas Instruments MSP430, Atmel Atmega128, and Intel px27ax families of microcontrollers. Currently, it does not support the family of ARM Cortex processors. The development of TinyOS has been discontinued since a few years.

ChibiOS/RT

ChibiOS/RT [41] is a compact and Open-Source RTOS. It is designed for embedded real-time applications where execution efficiency and compact code are important requirements. This RTOS is characterized by its high portability, compact size and, mainly, by its architecture optimized for extremely efficient context switching. It supports a preemptive thread model but it does not support stack sharing among threads.

The official list of supported microcontrollers is mainly focused on the ARM Cortex-M family, even if a very few other processors (i.e., ARM7, AVR Mega, MSP430, Power Architecture e200z, and STM8) are supported as well. Some further microcontrollers are not officially supported, and the porting of the RTOSs has been done by individual developers.

The footprint of this RTOS is very low, being between 1 KB and 5.5 KB.

ChibiOS/RT is provided under several licenses. Besides the commercial license, unstable releases are available as GPL v3 and stable releases as GPL v3 with linking exception. Since version 3 of GPL does not allow "tivoization" [42], these Open-Source licenses are not suitable for industrial contexts where the manufacturer wants to prevent users from running modified versions of the software through hardware restrictions.

ERIKA Enterprise v2

Erika Enterprise v2 [43] is a minimal RTOS providing hard real-time guarantees. It is developed by partner Evidence Srl, but it is released for free. The Open-Source license – GPL with linking exception (also known as "Classpath") [29] – is suitable for industrial usage because it allows linking (even statically) the proprietary application code with the RTOS without the need of releasing the source code.

The RTOS was born in 2002 to target the automotive market. During the course of the years it has been certified OSEK/VDX and it is currently used by either automotive companies (as Magneti Marelli and Cobra) or research institutions. ERIKA Enterprise v2 implements the AUTOSAR API 4.0.3 as well, up to Scalability Class 4.

Besides the very small footprint (about 2–4 KB), ERIKA Enterprise has innovative features, like advanced scheduling algorithms (e.g., resource reservation, immediate priority ceiling, etc.) and stack sharing to reduce memory usage.

It supports several microcontrollers (from 8-bit to 32-bit) and it has been one of the first RTOSs supporting multicore platforms (i.e., Altera NiosII). The current list of supported architectures includes Atmel AVR and Atmega, ARM 7 and Cortex-M, Altera NiosII, Freescale S12 and MPC, Infineon Aurix and Tricore, Lattice Mico32, Microchip dsPIC and PIC32, Renesas RX200, and TI MSP430. A preliminary support for ARM Cortex-A as well as the integration with Linux on the same multicore chip has been shown during a talk at the Automotive Linux Summit Fall [44] in October 2013.

Version 3 of ERIKA Enterprise has also been released recently [45]. The architecture of ERIKA Enterprise v3 has been directly derived as an evolution of the work described in this chapter, and is aimed to support full AUTOSAR OS compliance on various single and multi-/manycore platforms, including support for hypervisors.

7.2.3 Classification of Embedded RTOSs

The existing open-source RTOSs can be grouped in the following classes:

1. POSIX RTOSs, which provide the typical POSIX API allowing dynamic thread creation and resource allocation. These RTOSs have a large footprint due to the implementation of the powerful but complex POSIX API. Examples are: Marte OS and RTEMS.
2. Simil-POSIX RTOSs, which try to offer an API with the same capabilities of POSIX (i.e., dynamic thread creation and resource allocation) but at a lower footprint meeting the typical constraints of small embedded systems. Examples are: FreeRTOS, Ecos and ChibiOS/RT.
3. OSEK RTOSs, implementing the OSEK/VDX API with static thread creation but still allowing thread preemption. These RTOSs are characterized by a low footprint. Moreover, they usually also offer stack-sharing among the threads, allowing the reduction of memory consumption at run-time. Examples are: ERIKA Enterprise, Trampoline, and FreeOSEK.
4. Other minimal RTOSs, which have a low footprint and a non-preemptive thread model by construction. Usually, these RTOSs offer the stack-sharing capability. Examples are: TinyOS and Contiki.

This classification is shown in the following Table 7.1:

Table 7.1   Classification of RTOSs

	POSIX	Simil-POSIX	OSEK	Other Minimal
API	POSIX	Custom	OSEK/VDX	Custom
Footprint size	Big	Medium	Small	Small
Thread preemption	V	V	V	X
Thread creation	V	V	–	–
Stack sharing	–	–	V	V
Examples	MarteOS	FreeRTOS	ERIKA Enterprise	TinyOS
	RTEMS	Ecos	Trampoline	Contiki
		ChibiOS/RT	FreeOSEK	QP

7.3 Requirements for The Choice of The Run Time System

This section includes a short description of the main requirements that influenced the choice of the OS platform for the implementation of our parallel programming model.

7.3.1 Programming Model

The run-time system is a fundamental software component of the parallel programming model to transform the parallel expressions defined by the user into threads that execute in the different processing units, i.e., cores.

Therefore, the OS system must provide support to execute the run-time system that will implement the API services defined by the parallel programming model. In our case, the requirement is related to the fact that an UNIX environment such as Linux should be present, with support for the C and C++ programming languages.

7.3.2 Preemption Support

In single-core real-time systems, allowing a thread to be preempted has a positive impact on the schedulability of the system because the blocking on higher-priority jobs is significantly limited. However, in many-core systems, the impact of preemptions on schedulability is not as clear, since higher priority jobs might have a chance to execute on one of the many other cores available in the system. Nevertheless, for highly parallel workloads, it may happen that all cores are occupied by lower-priority parallel jobs, so that higher-priority instances may be blocked for the whole duration of the lower-priority jobs. In this case, a smart preemption support might be beneficial, allowing a subset of the lower-priority instances to be preempted in favor of the higher-priority jobs. The remaining lower-priority instance may continue executing on the remaining cores, while the state of the preempted instances needs to be saved by the OS, in order to restore it as soon as there are computing units available again.

In order to develop the proper OS mechanisms, it is necessary to support the kind of preemption needed by the scheduling algorithms described in Chapter 4, with particular reference to the hybrid approach known as "limited preemption," and to the store location of the preempted threads context. In order to implement such techniques, the OS design needs to take into account which restrictions will be imposed on the preemptability of the threads, whether by means of statically defined preemption points, or by postponing the invocation of the scheduling routine by a given amount of time.

7.3.3 Migration Support

In migration-based multicore systems, a preempted thread may resume its execution on a different core. Migration support requires additional OS mechanisms to allow threads to be resumed on different cores. Different migration approaches are possible:

- Partitioned approach: Each thread is scheduled on one core and cannot execute on other cores;
- Clustered approach: Each thread can execute on a subset (cluster) of the available cores;
- Global approach: Threads can execute on any of the available cores.

7.3.4 Scheduling Characteristics

Real-time scheduling algorithms are often divided into static vs. dynamic scheduling algorithms, depending on the priority assigned to each job to execute. Static algorithms assign a fixed priority to each thread. Although they are easier to implement, their performance could be lower than with more flexible approaches that may dynamically change priorities of each thread. Depending on the scheduling strategy, fixed or dynamic, different OS kernel mechanisms will be needed.

Another design point concerns the policies for arbitrating the access to mutually exclusive shared resources. Depending on the adopted policy, particular synchronization mechanisms, thread queues, and blocking primitives may be needed.

7.3.5 Timing Analysis

In order for the timing analysis tools to be able to compute safe and accurate worst-case timing estimates, it is essential that the RTOS manages all the software/hardware resources in a predictable manner. Also, it is crucial for the timing estimates to be as tight as possible because subsequently these values (like the worst-case execution time of a task or the maximum time to read/write data from/to the main memory) will propagate all the way up and will be used as basic blocks in higher-level analyses like the schedulability analysis. Deriving tight estimates requires that all the OS mechanisms that allocate and arbitrate the access to the system resources are thoroughly documented and do not make use of any procedure that involves randomness or based on non-deterministic parameters.

Task-to-thread and thread-to-core mapping: The allocation of the tasks to the threads and the mapping of the threads to the cores must be documented; ideally, it should also be static and known at design time. If the allocation is dynamic, i.e., computed at run-time, then the allocation/scheduling algorithm should follow a set of deterministic (and fully documented) rules. The knowledge of where and when the tasks execute considerably facilitates the timing analysis process, as it allows for deriving an accurate utilization profile of each resource and then uses those profiles to compute safe bounds on the time it takes to access these resources.

Contract-based resource allocation scheme: Before executing, each application or task has a "contract" with every system resource that it may need to access. Each contract stipulates the minimum share of the system resource (hardware and software) that the task must be allowed to use over time. Considering a communication bus shared between several tasks, a TDMA (Time Division Multiple Access) bus arbitration policy is a good example of a contract-based allocation scheme: the number of time-slots dedicated to each task in a time-frame of fixed length gives the minimum share of the bus that is guaranteed to be granted to the task at run-time. When the resource is a core, contract-based mechanisms are often referred to as reservation-based scheduling. Before executing, an execution budget is assigned to every task and a task can execute on a core only if its allocated budget is not exhausted. Technically speaking, within such reservation-based mechanisms, the scheduling algorithm of the OS does not schedule the execution of the tasks as such, but rather it manages the associated budgets (i.e., empties and replenishes them) and defines the order in which those budgets are granted to the tasks. There are many advantages of using contract-based mechanisms. For example, they provide a simple way of protecting the system against a violation of the timing parameters. If a task fails and starts looping infinitely, for instance, the task will eventually be interrupted once it runs out of budget, without affecting the planned execution of the next tasks. These budgets/contracts can be seen as fault containers. They guarantee a minimum service to every task while enabling the system to identify potential task failure and avoid propagating the potentially harmful consequences of a faulty task through the execution of the other tasks.

Runtime budget/contract reinforcement: Mechanisms must be provided to force the system resources and the tasks to abide with their contract, e.g., a task is not allowed to execute if its CPU budget is exhausted or if its budget is not currently given to that task by the scheduler. This mechanism is known in the real-time literature as "hard reservation."

Memory isolation: The OS should also provide mechanisms to dedicate regions of the memory to a specific task, or at least to tasks running on a specific core.

Execution independence: The programs on each core shall run independent of the hardware state of other cores.

7.4 RTOS Selection

Considering the architecture of the reference platform (i.e., host processor connected to a set of accelerators, similarly to other commercially available many-core platforms), we decided to use two different OSs for the host and the many-core processors.

7.4.1 Host Processor

Linux has been chosen for the host processor, due to its excellent support for peripherals and communication protocols, the several programming models supported, and the popularity in the HPC domain.

Given the nature of the project and the requirements of the use-cases, soft real-time support has been added through the adoption of the PREEMPT_RT patch [10].

7.4.2 Manycore Processor

For the manycore processor, a proper RTOS was needed. The selected RTOS should have been Open-Source and lightweight (i.e., with a small footprint) but providing a preemptive thread model. For these reasons, only the RTOSs belonging to columns 2 (i.e., Simil-POSIX) and 3 (i.e., OSEK) of Table 7.1 could be selected. Moreover, the selected RTOS must be actively maintained through the support of a development community.

Ecos has been discarded due to the big footprint (comparable to the one of POSIX systems). FreeOSEK has been discarded because the project is not actively maintained and because it does not offer any additional feature with respect to ERIKA Enterprise. ChibiOS/RT, Trampoline, and QP, instead, have been discarded for the too restrictive open-source license, not suitable for industrial products.

The only RTOSs that fulfilled our requirements, therefore, were ERIKA Enterprise [43] and FreeRTOS [28]. The project eventually chose to use ERIKA Enterprise due to its smaller footprint, the availability of advanced real-time features, and the strong know-how in the development team.

7.5 Operating System Support

7.5.1 Linux

As for the Linux support, we started with the Linux version provided together with the reference platform. In particular, the Kalray Bostan AccessCore SDK included an Embedded Linux version 3.10, and on top of it we assembled and configured a filesystem based on the Busybox project [46] produced using Buildroot [47].

The Linux version provided included Symmetric Multi-Processing (SMP) support (which is a strong requirement for running PREEMPT_RT [10]), and included the PREEMPT_RT patch.

7.5.2 ERIKA Enterprise Support

We have successfully ported the ERIKA Enterprise [43] on the MPPA architecture, supporting its VLIW (Very Large Instruction Word) Instruction Set Architecture (ISA) and implementing the API used by the off-loading mechanism. The following paragraphs list the main challenges we had during the porting, and the main choices we addressed, together with some early performance results.

7.5.2.1 Exokernel support

The development on the platform directly supports the Kalray "exokernel," which is a set of software, mostly running on the 17th core of each cluster (the resource manager core), used to provide a set of services needed to let a cluster appear "more like" a SMP machine. Among the various services, the exokernel includes communication services and inter-core interrupts. The exokernel API is guaranteed to be maintained across chip releases, while the raw support for the resource manager core will likely change with newer chip releases.

7.5.2.2 Single-ELF multicore ERIKA Enterprise

One of the main objectives during the porting of the ERIKA RTOS has been the reduction of the memory footprint of the kernel, obtained by using a Single-ELF build system.

The reason is that the multicore support in ERIKA was historically designed for hardware architectures which did not have a uniform memory region, such as Janus [48]. In those architectures, each core had its own local memory and, most importantly, the view of the memory as seen by the various cores was different (that is, the same memory bank was available at a different address on each core). This imposed the need for a custom copy of the RTOS for each core. Other architectures had a uniform memory space, but the visibility of some memory regions was prevented by the Network on Chip. On Altera Nios II, for example, addresses differentiating by only the 31st bit referred to the same physical address with or without caching. This, again, implied the need for separate images (in particular, you can refer to the work done during the FP6 project FRESCOR, D-EP7 [49]). More modern architectures like Freescale PPC and Tricore AURIX allowed the possibility of single-ELF, but the current multi-ELF scaled relatively well on a small number of cores, reducing the need for single-ELF versions of the system.

In manycore architectures such as Kalray, the multi-ELF approach showed its drawback: the high number of cores, in fact, required avoiding code duplication to not waste memory. Moreover, each core has the visibility of a memory region, and the addressing is uniform across the cores. For this reason, after an initial simple single-core port of ERIKA on the Kalray MPPA, the project decided to eventually design a single-ELF implementation; this activity required a complete rewrite of the codebase (named ERIKA Enterprise v3). The new codebase is now in production and sponsored through a dedicated website [45] in order to gather additional comments and feedbacks. The next paragraphs include a short description of the main design guidelines, which are also described in a specific public document [50].

7.5.2.3 Support for limited preemption, job, and global scheduling

The ERIKA Enterprise RTOS traditionally supported partitioned scheduling, where each core has a set of statically assigned tasks which can be individually activated.

In order to support the features requested by the parallel programming framework, the ERIKA Enterprise scheduler has been modified to allow the following additional features:

- **Limited preemption scheduler** – ERIKA Enterprise has been improved to allow preemptions only at given instants of time (i.e., at task scheduling points, see Chapters 3 and 6). The main advantage is related to performance, because the preemption is implemented in a moment that has a limited performance hit on the system.
- **Job activation** – In ERIKA Enterprise, each task can be individually activated as the effect of an *ActivateTask* primitive. In the new environment, the OS tasks are mapped onto the OpenMP worker threads (see Chapter 6). Those threads are activated in "groups" (named here "jobs"), because their activation is equivalent to the start of an OpenMP offload composed by N OS tasks on a cluster. For this reason, ERIKA Enterprise now supports "Job activation," which allows activating a number of tasks on a cluster. Typically, those tasks will have all the same priority (as they map the execution of an OpenMP offload).
- **Global scheduling** – In order to obtain the maximum throughput, ERIKA implemented a work conserving global scheduler, which is able to implement migration of tasks among cores of the same cluster. The migration support also handles contention on the global queue in case there are two or more cores idle.

7.5.2.4 New ERIKA Enterprise primitives

The implementation of ERIKA Enterprise required the creation of a set of *ad hoc* primitives, which have been included in a new kernel explicitly developed for Kalray. The new primitives are described below:

CreateJob: This primitive is used to create a pool of OS tasks which are coordinated for the parallel execution in a cluster. A "Job" is composed by a maximum number of tasks which is equal to the cluster size (16 on Kalray MPPA). It is possible to specify how many tasks should be created, and on which cores they should be mapped in case of partitioned scheduling. All tasks which are part of a Job have the same priority, the same entry point, the same stack size. Finally, they all have an additional parameter which is used by the OpenMP workers to perform their job.

ReadyJob and **ActivateJob:** These two primitives are used to put in the ready queue (either global or partitioned depending on the kernel configuration) the tasks corresponding to a specific mask passed as parameter (the mask is a subset of the one passed previously to CreateJob). In addition to this, ActivateJob adds a preemption point on the calling site and issues inter-core interrupts in full preemptive configuration.

JoinJob: This is a synchronization point at the termination of all tasks of a Job. It must be called on a task which has lower priority than the Job task priority.

Synchronization primitives are also provided to allow the implementation of use-level locks and higher-level programming model synchronization constructs for the OpenMP runtime library (discussed in Chapter 6).

SpinLockObj and **SpinUnlockObj:** These primitives provide a standard lock API, and are directly based on spinlock primitives provided by the Kalray HAL. At the lowest abstraction level, the lock data structure is implemented as a 32-bit integer, which could be allocated at any memory-mapped address. Using this approach, the lock variables can be statically allocated whenever it is possible, and when more dynamism is required, lock data structures can be initialized via standard malloc operations on a suitable memory range.

WaitCondition and **SignalValue:** These primitives provide a synchronization mechanism based on WAIT/SIGNAL semantics. ERIKA supports four condition operators (equal, not equal, lower than, greater than) and three different wait policies:

1. BLOCK_NO – The condition is checked in a busy waiting loop;
2. BLOCK_IMMEDIATELY – The condition is checked once. If the check fails (and no other tasks are available for execution) the processor enters sleep mode until the condition is reached. A specific signal is then used to wake-up the processor.
3. BLOCK_OS – Informs the OS that the ERIKA task (i.e., the OpenMP thread mapped to that task) is voluntarily yielding the processor. The OS can then use this information to implement different scheduling policies. For example, the task can be suspended and a different task (belonging to a different job) can be scheduled for execution.

7.5.2.5 New data structures

Addressing the single-ELF image implementation in the end required a restructuring of the kernel data structures.

The initial version of ERIKA Enterprise used a set of global data structures (basically, C arrays of scalars) allocated in RAM or ROM. Each core had its own copy of the data structures, with the same name. Data which is shared among the cores is defined and initialized in one core referred to as the *master* core. The other cores are called *slave* cores. Afterwards, when compiling the *slave* cores' code, the locations of the shared data are appended to each core's linker scripts (see also [48]). Figure 7.2 shows the structure of the two ELF files, highlighting the first core (*master*), which has everything defined, and the subsequent *slave* cores, which have the shared symbols addresses appended in the linker script.

The single-ELF approach required a complete restructuring of the binary image. The complete system is compiled in a single binary image, and the data structures are designed to let the cores access the relevant per-CPU data. The main guidelines used when designing the data structures are the following:

- All data is shared among all cores.
- The code must be able to know on which core it is running. This is done typically using a special register of the architecture that holds the CPU number.
- Given the CPU number, it is possible to access "private" data structures to each core (see Figure 7.3). Note that those "private" data structures can be allocated in special memory regions "near" each core (for example, they could be allocated in sections which can be pinned to per-core caches).
- Clear distinction between Flash Descriptor Blocks (named *DB) and RAM Control Blocks (named *CB). In this way the reader has a clear idea of the kind of content from the name of the data structure.
- Limited usage of pointers (used to point only from Flash to RAM), to make the certification process easier.

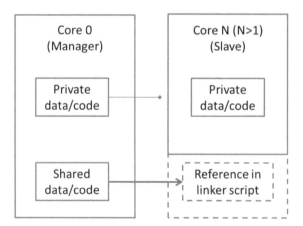

Figure 7.2 Structure of the multicore images in the original ERIKA Enterprise structure.

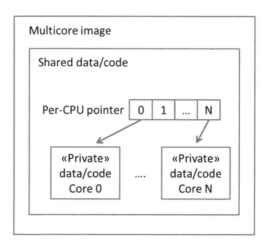

Figure 7.3   Structure of the Single-ELF image produced by ERIKA Enterprise.

7.5.2.6 Dynamic task creation

In the original version of ERIKA, RTOS tasks were statically allocated by defining them inside an OIL file. In the new version of ERIKA, we allowed a pre-allocation of a given number of RTOS tasks, which can be afterwards "allocated" using a task creation primitive. In this way, the integration with the upper layers of OpenMP becomes simpler, as OpenMP makes the hypothesis of being able to create as many threads as needed using the underlying Linux primitive pthread_create.

In addition to the changes illustrated above, we also took the opportunity for making the following additional changes to ease future developments.

7.5.2.7 IRQ handlers as tasks

The original version of ERIKA handled interrupts in the most efficient way in the case of no memory protection among tasks. When memory protection comes into play, treating IRQs as special tasks has the advantage of simplifying the codebase.

In view of the future availability of multi-many cores with memory protection we implemented the possibility for an IRQ to be treated as a task. A special fast handler is called upon IRQ arrival, which has the main job of activating the "interrupt task."

This approach also simplified the codebase by allowing a simpler context change primitive, which in turn simplifies the implementation in VLIW chips such as Kalray.

7.5.2.8 File hierarchy

For the new version of ERIKA, we adopted a new file hierarchy which aims to a simplification of the codebase. In particular, the main changes of the new codebase are the following:

- In the old version, CPU (the specific instruction set, such as PPC, Cortex-MX, etc.), MCU (the peripherals available on a specific part number), Boards (code related to the connections on the PCB) were stored in directories under the "pkg" directory. With the growing number of architectures supported, this became a limitation which also made the compilation process longer. The new version of the codebase includes MCUs and Boards under the CPU layer, making the dependencies in the codebase clearer.
- We adopted a local self-contained flat (single directory) project structure instead of a complex hierarchy. All needed files are copied once in the project directory at compilation time, leading to simpler makefiles.
- We maintained the RTOS code separated from the Application configuration. This is very useful to allow the deployment of pre-compiled libraries; moreover it allows partial compilation of the code.

7.5.2.9 Early performance estimation

Before implementing the Single-ELF version of ERIKA on Kalray, we performed an initial implementation of the traditional single-core porting of ERIKA in order to get a reference for the evaluation of the subsequent development. Please note that the evaluation of the new version of ERIKA has been done on a prototype implementation (not the final one). However, the numbers are good enough to allow a fair comparison of the two solutions.

Table 7.2 summarizes an early comparison between the old and the new implementation of ERIKA, for a simple application with two tasks on a single core. The purpose of the various columns is the following:

- The comparison between the second and the third column gives a rough idea of the difference in the ISA on a "reasonably similar" code on another (different) architecture, Nios II.
- The comparison between the third and the fourth column gives a rough idea of the impact of the changes of the new version of ERIKA over the old version. The values show an increase of the code footprint. This increase, however, is less than indicated by the table: the old version of ERIKA, in fact, does contain the support for multiple task activations (which has not been compiled) and dynamic task creation (which was

Table 7.2 ERIKA Enterprise footprint (expressed in bytes)

Description	Old Version (*)	Old Version	New Version Single-core	New Version, Multicore with Services for Supporting Libgomp
Platform	Nios II	Kalray MPPA	Kalray MPPA	Kalray MPPA
Code footprint (**)	About 800	1984	2940	4156[1]
Code footprint related to multicore (***)	–	–	–	4156 + 502 for RM
Flash/Read-only	164	–	–	–
RAM		192	216	216 + 128 for each core[2]

(*) Numbers taken from D-EP7v2 of the FRESCOR project [49]. These numbers can be taken as a reference for the order of magnitude for the size of the kernel and may not represent the same data structures. We considered these numbers as the current implementation on ERIKA has roughly a similar size and they can be used as a reference for comparing "similar" implementations.
(**) The code footprint includes the equivalent of the following functions: *StartOs*, *ActivateTask*, *TerminateTask*
(***) Code related to the handling of the multicore features (remote notifications, inter-processor interrupts, spin locks, and code residing) on the Resource Manager Core (see Chapter 2).

not available). Moreover, we have to consider that the old version of ERIKA needed 1,984 bytes for each core. The new version of ERIKA, instead, needs 2,940 bytes, regardless of the number of cores. This means that with just two cores, the amount of memory needed by the new version of ERIKA Enterprise is less than using the old version of the RTOS.

• The comparison between the fourth and the fifth column gives a rough idea of the impact of the multicore support. The increase of the code footprint is mainly due to additional synchronization primitives (i.e., spinlocks) needed for distributed scheduling – i.e., to allow the "group activation" done by the Resource Manager on behalf of OpenMP. Therefore, this increase is specific to the Kalray architecture, and it is missing on other (e.g., shared-memory) architectures. Note that the footprint takes into account only the kernel part with the services for supporting the OpenMP runtime library; it does not include the library itself.

---

[1]The footprint takes into account only kernel and support for the OpenMP runtime library; it does not include the library itself.
[2]$128 = 44$ (core data structures) + 84 (idle task).

Table 7.3 Timings (expressed in clock ticks)

Feature	Time on ERIKA
ActivateTask, no preemption	384
ActivateTask, preemption	622
An IRQ happens, no preemption	585
An IRQ happens, with preemption	866

Table 7.3 provides basic measurements of activation and pre-emption of tasks on a single-core:

Tables 7.4 and 7.5 provide some timing references to compare ERIKA Enterprise (which is a RTOS) with NodeOS on MPPA-256, taken using the Kalray MPPA tracer. Since NodeOS does not support preemption (and therefore a core can execute only one thread) we have configured ERIKA Enterprise to run only one task on each core as well. Then, we have measured footprint and execution times. In particular, Table 7.4 provides a rough comparison of the footprint for ERIKA and NodeOS on Kalray MPPA. For ERIKA, the footprint also takes into account the per-core and per-task data structures in a cluster composed of 17 cores. This footprint can be reduced by using a static configuration of the RTOS. Table 7.5 provides a comparison between the thread creation time on NodeOS and the equivalent inter-core task activation on ERIKA.

Table 7.4 Footprint comparison between ERIKA and NodeOS for a 16-core cluster (expressed in bytes)

Description	ERIKA New Version, Multicore with Services for Supporting OpenMP	NodeOS
Code footprint	4484[3]	10060
RAM	2184	2196

Table 7.5 Thread creation/activation times (expressed in clock ticks)

Inter-core Task Activation on ERIKA	Thread Creation on NodeOS
1200	3300

---

[3]The footprint takes into account only kernel and support for libgomp; it does not include the whole libgomp library.

7.6 Summary

This chapter illustrated the state of the art of the OSs suitable for the reference parallel programming model. After reviewing the main requirements that influenced the implementation, the selection of the RTOS for the reference platform has been described for both the host processor and the manycore accelerators. Furthermore, a description of the main implementation choices for the ERIKA Enterprise v3 and Linux OS have been detailed. As can be seen, the result of the implementation provides a complete system which is capable of addressing high-performance workloads thanks to the synergies between the general-purpose OS Linux and the ERIKA Enterprise RTOS.

References

[1] *Top500, Linux OS*. Available at: http://www.top500.org/statistics/details/osfam/1

[2] *RCU*, available at: http://en.wikipedia.org/wiki/Read-copy-update

[3] *GNU General Public License (GPL)*, available at: https://www.gnu.org/copyleft/gpl.html

[4] Lipari, G., Scordino, C., *Linux and Real-Time: Current Approaches and Future Opportunities, International Congress ANIPLA*, Rome, Italy, 2006.

[5] *RTLinux*, available at: http://en.wikipedia.org/wiki/RTLinux

[6] *RTAI – the RealTime Application Interface for Linux*, available at: https://www.rtai.org/

[7] Xenomai, *Real-Time Framework for Linux*. Available at: http://www.xenomai.org/

[8] Windriver, *VxWorks RTOS*. Available at: http://www.windriver.com/products/vxworks/

[9] *POSIX IEEE standard*, available at: http://en.wikipedia. org/wiki/POSIX

[10] *The Real Time Linux project*, available at: https://wiki.linuxfoundation.org/realtime/start

[11] *The Linux Foundation*, available at: https://www.linuxfoundation.org/

[12] Libenzi, D., *SCHED SOFTRR Linux Scheduler Policy*, available at: http://xmailserver.org/linux-patches/softrr.html

[13] Kolivas, C., *Isochronous class for unprivileged soft RT scheduling*. Available at: http://ck.kolivas.org/patches/

[14] *SCHED_DEADLINE Linux Patch*, available at: http://en.wikipedia.org/wiki/SCHED_DEADLINE

[15] Lelli, J., Scordino, C., Abeni, L., Faggioli, D., Deadline scheduling in the Linux kernel, Software: Practice and Experience, 46, pp. 821–839, 2016.

[16] *ACTORS European Project*, available at: http://www.actors-project.eu/

[17] *Earliest Deadline First (EDF)*, available at: http://en.wikipedia.org/wiki/Earliest_deadline_first_scheduling

[18] Green Hills, *Integrity RTOS*. Available at: http://www.ghs.com/products/rtos/integrity.html

[19] *QNX RTOS*, available at: http://www.qnx.com/

[20] *SYSGO PikeOS*, available at: http://www.sysgo.com/products/pikeos-rtos-and-virtualization-concept/

[21] Mentor Graphics, *Nucleus RTOS*. Available at: http://www.mentor.com/embedded-software/nucleus/

[22] *LynuxWorks LynxOS*, available at: http://www.lynuxworks.com/rtos/

[23] *Micrium µc/OS-III*, available at: http://micrium.com/

[24] *Segger EmbOS*, available at: http://www.segger.com/embos.html

[25] *ENEA OSE*, available at: http://www.enea.com/ose

[26] *Arccore Arctic Core*, available at: http://www.arccore.com/products/

[27] Wikipedia, *List of Real-Time Operating Systems*, available at: http://en.wikipedia.org/wiki/List_of_real-time_operating_systems

[28] *FreeRTOS*, available at: http://www.freertos.org/

[29] *GPL Linking Exception*, available at: http://en.wikipedia.org/wiki/GPL_linking_exception

[30] *Contiki*, available at: http://www.contiki-os.org/

[31] Wikipedia, *Protothreads*. Available at: http://en.wikipedia.org/wiki/Protothreads

[32] *Marte OS*, available at: http://marte.unican.es/

[33] *Ecos RTOS*, available at: http://ecos.sourceware.org/

[34] *FreeOSEK RTOS*, available at: http://opensek.sourceforge.net/

[35] Quantum Leaps, QP^{TM} *Active Object Frameworks for Embedded Systems*, available at: http://www.state-machine.com/

[36] *Trampoline RTOS*, available at: http://trampoline.rts-software.org/

[37] *RTEMS*, available at: http://www.rtems.org/

[38] *Footprint of RTEMS*, available at: http://www.rtems.org/ml/rtems-users/2004/september/msg00188.html

[39] *Tiny RTEMS*, available at: https://code.google.com/p/tiny-rtems/

[40] *TinyOS*, available at: http://www.tinyos.net/

[41] *ChibiOS/RT*, available at: http://www.chibios.org/

[42] *Tivoization*, available at: http://en.wikipedia.org/wiki/Tivoization

[43] *Erika Enterprise RTOS*, available at: http://erika.tuxfamily.org

[44] *Automotive Linux Summit Fall*, available at: http://events.linuxfounda tion.org/events/automotive-linux-summit-fall

[45] *ERIKA Enterprise v3*, available at: http://www.erika-enterprise.com

[46] *Busybox*, available at http://www.busybox.net/, last accessed March 2016.

[47] *Buildroot*, available at http://buildroot.uclibc.org/, last accessed March 2016.

[48] Ferrari, A., Garue, S., Peri, M., Pezzini, S., Valsecchi, L., Andretta, F., and Nesci, W., "The design and implementation of a dual-core platform for power-train systems." In *Convergence 2000*, Detroit, MI, USA, 2000.

[49] *FRESCOR FP6 D-EP7v2*, available at http://www.frescor.org/ and also http://www.frescor.org/ and also http://erika.tuxfamily.org/wiki/index. php?title=Altera_Nios_II, last accessed March 2016.

[50] Evidence, *ERIKA Enterprise Version 3 Requirement Document*, available at ERIKA Enterprise website: http://erika.tuxfamily.org/drupal/ content/erika-enterprise-3

Index

About the Editors

Luís Miguel Pinho is Professor at the Department of Computer Engineering of the School of Engineering, Polytechnic Institute of Porto, Portugal, with a PhD in Electrical and Computer Engineering at the University of Porto, Portugal. He has more than 20 years of experience in research in the area of real-time and embedded systems, particularly in concurrent and parallel programming models, languages, and runtime systems. He is Research Associate in the CISTER research unit, where he was Vice-Director from 2010 to 2017, being responsible for creating several research areas and topics, among which the activities on parallel real-time systems, that he leads. He has participated in more than 20 R&D projects, was Project Coordinator and Technical Manager of the FP7 R&D European Project P-SOCRATES and national-funded CooperatES and Reflect Projects. He was also coordinator of the participation of CISTER and work package leader in several other international and national projects. He has published more than 100 papers in international conferences and journals in the area of real-time embedded systems. He was Senior Researcher of the ArtistDesign NoE and is currently a member of the HiPEAC NoE. He was Keynote Speaker at the 16th IEEE Conference on Embedded and Real-Time Computing Systems and Applications (RTCSA 2010) and is the Editor-in-Chief of the Ada User Journal. Among others, he was General Co-Chair of the 28th GI/ITG International Conference on Architecture of Computing Systems (ARCS 2015), and Program Co-Chair of the 24th International Conference on Real-Time Networks and Systems (RTNS 2016) and of the 21st International Conference on Reliable Software Technologies (Ada-Europe 2016).

Eduardo Quiñones is a senior researcher in the group on Interaction between the Computer Architecture and the Operating System (CAOS) at BSC and member of HIPEAC. He worked at the Intel Barcelona Research Center from 2002 till 2004 in compiler techniques for EPIC architectures (including Itanium I and II). At BSC, he has previous experiences involved in the architectural definition and the avionics case study definition in the MERASA

FP7 project and he leads the architectural definition work packages of the PROARTIS and the parMERASA FP7 projects, and lead the applicability of HPC parallel programming models to real-time embedded systems to increase performance in the P-SOCRATES FP7 project. Moreover, he is involved in two research projects with the European Space Agency (ESA), one as a technical manager. His research area focuses on compiler techniques and many-core architectures for safety-critical systems on which he is co-advising six PhD students. He is currently the project coordinator for the CLASS H2020 project.

Marko Bertogna is Associate Professor at the University of Modena (Italy), where he leads the High-Perfomance Real-Time Systems Laboratory (HiPeRT Lab). His main research interests are in High-Performance Real-Time systems, especially based on multi- and many-core devices, Autonomous Driving and Industrial Automation systems, with particular relation to related timing and safety requirements. Previously, he was Assistant Professor at the Scuola Superiore Sant'Anna of Pisa, working at the Real-Time Systems Lab since 2003. He graduated magna cum laude in Telecommunication Engineering at the University of Bologna in 2002. From 2001 to 2002, he worked on integrated optical devices at the Technical University of Delft, The Netherlands. In 2006, he visited the University of North Carolina at Chapel Hill, working with prof. Sanjoy Baruah on scheduling algorithms for single and multicore real-time systems. In 2008, he received a PhD in Computer Sciences from the Scuola Superiore Sant'Anna of Pisa, with a dissertation on Real-Time Systems for Multicore Platforms, awarded as the best scientific PhD thesis discussed at Scuola Superiore Sant'Anna in 2008 and 2009.

Andrea Marongiu received the PhD degree in electronic engineering from the University of Bologna, Italy, in 2010. He has been a postdoctoral reserch fellow at ETH Zurich, Switzerland. He currently holds an assistant professor position at the University of Bologna (Department of Computer Science and Engineering). His research interests focus on programming models and architectures in the domain of heterogeneous multi- and many-core systems on a chip. This includes language, compiler, runtime and architecture support to efficiently address performance, predictability, energy and reliability issues in paralle, embedded systems, as well as HW-SW co-design of accelerator-based MPSoCs. In this field, he has published more than 100 papers in international peer-reviewed conferences and journals, with more than 700 citations and an h-index of 16 [Google Scholar]. He has collaborated with several international research institutes and companies.

Vincent Nélis earned his PhD degree in Computer Science at the University of Brussels (ULB) in 2010. Since then, he has been working at CISTER as a Research Associate. He is an expert in real-time scheduling theory with a focus on multiprocessor/multicore systems and in interference analysis, including pre-emption cost analysis and bus/network contention analysis in multicores and many-cores systems. Vincent is regularly a member of technical program committees for international conferences, workshops, and journals. He has graduated 2 PhDs and he is currently the supervisor of a third PhD student. He has contributed to 5 R&D projects and published 25+ papers with about 30 different co-authors in international conferences and scientific journals. His work was awarded at several occasions: "Solvay Award" (2006), "Outstanding Paper Award" (2012), two "Best Paper Awards" (2010 and 2013) and a "Best Presentation Award" (2013).

Paolo Gai graduated (cum laude) in Computer Engineering at University of Pisa in 2000. He obtained the PhD from Scuola Superiore Sant'Anna in 2004. Since 2002 he is founder of Evidence Srl, a company providing innovations in the field of operating systems and platforms for embedded devices in the automotive and industrial fields.

His research activity is focused on the development of hard real-time architectures for embedded automotive control systems. His research interests include multi and many-core processor systems, object-oriented programming, real-time operating systems, scheduling algorithms, multimedia applications, and hypervisors.

Juan Sancho holds a degree in Telecommunications Engineering from the Universidad Politécnica de Valencia, Spain. He developed his Final Project Degree in the field of Health Monitoring using Wireless Sensor Networks at the Wireless Centre of the Copenhagen University of Engineering, Denmark. In the past he worked as network & systems engineer, participating in several European FP7, ENIAC and National projects (BUTLER, TOISE, SICRA, TSMART). Since 2014 he works as a Research & Innovation Engineer in ATOS Research & Innovation division, collaborating in FP7 and H2020 projects related to IoT topics (COSMOS, P-SOCRATES) and the Energy domain (inteGRIDy, ELVITEN, eDREAM). His research interests cover Big Data & Edge platforms, DevOps, Renewable Energy Sources, WSN and low-power embedded systems.